YANMAR MARINE DIESEL ENGINE
3YM30/3YM20/2YM15

SERVICE MANUAL

YANMAR MARINE DIESEL ENGINE 3YM30/3YM20/2YM15

SERVICE MANUAL

ISBN/EAN: 9783954275083
Erscheinungsjahr: 2012
Erscheinungsort: Bremen, Deutschland

© maritimepress in Europäischer Hochschulverlag GmbH & Co. KG, Fahrenheitstr. 1, 28359 Bremen. Alle Rechte beim Verlag und bei den jeweiligen Lizenzgebern.

www.maritimepress.de | office@maritimepress.de

Bei diesem Titel handelt es sich um den Nachdruck eines historischen, lange vergriffenen Buches. Da elektronische Druckvorlagen für diese Titel nicht existieren, musste auf alte Vorlagen zurückgegriffen werden. Hieraus zwangsläufig resultierende Qualitätsverluste bitten wir zu entschuldigen.

FOREWORD

This service manual has been complied for engineers engaged in sales, service, inspection and maintenance. Accordingly, descriptions of the construction and functions of the engine are emphasized in this manual, while items, which should already be common knowledge, are omitted.

One characteristic of a marine diesel engine is that its performance in a vessel is governed by the applicability of the vessel's hull construction and its steering system.

Engine installation, fitting out and propeller selection have a substantial effect on the performance of the engine and the vessel. Moreover, when the engine runs unevenly or when trouble occurs, it is essential to check a wide range of operating conditions - such as installation to the full and suitability of the ship's piping and propeller - and not just the engine itself. To get maximum performance from this engine, you should completely understand its functions, construction and capabilities, as well as proper use and servicing.

Use this manual as a handy reference in daily inspection and maintenance, and as a text for engineering guidance.

Model 3YM30 has been used for the illustrations in this service manual, but they apply to other models in the 3YM series engines.

California Proposition 65 Warning

Diesel engine exhaust and some of its constituents are known to the State of California to cause cancer, birth defects, and other reproductive harm.

California Proposition 65 Warning

Battery posts, terminals, and related accessories contain lead and lead compounds, chemicals known to the State of California to cause cancer and reproductive harm.
Wash hands, after handling.

©2005 YANMAR CO., LTD
All rights reserved. This manual may not be reproduced or copied, in whole or in part, without the written permission of YANMAR CO., LTD.

Publication No.	0BYMM-G00100

History of Revision

Manual Name	Service Manual for Marine Diesel Engine
Engine Model:	3YM30/3YM20/2YM15

Number of revision	Date of revision	Reason for correction	Outline of correction	Correction item No (page)	Corrected by
New edition	March 2004				
R1	April 25, 2005	Add : Eug. Model 2YM15	1. Specifications table added. 2. Engine outline added. 3. Piping diagram added. 4. Other data added.		Quality Control Dept. Marine Factory
		Corrected data. (3YM30, 3YM20)	1. Front cover added to engine outline. 2. Notice on seawater pump rotating direction was changed. 3. Incorrect word corrected.	P6-P9 P32	
R2	Nov. 2006	Corrected data.	1. Engine mounting bolt side pitch (3YM30 with KM2P-1). 2. Engine mounting bolt side pitch (3YM30C with SD20). 3. No-load max. speed (3YM20).	P6 P7 P44	YMI Business Development Department

Printed in Japan

CONTENTS

1. General 1
   1.1 Exterior views 1
   1.2 Specifications 2
   1.3 Fuel oil, lubricating oil and cooling water 4
      1.3.1 Fuel oil 4
      1.3.2 Lubricating oil 4
      1.3.3 Cooling water 5
   1.4 Engine outline 6
   1.5 Piping diagrams 10
   1.6 Exhaust gas emission regulation 12
      1.6.1 Engine identification (EPA/ARB) 12
      1.6.2 Exhaust gas emission standard (EPA/ARB) 13
      1.6.3 Guarantee conditions for emission standard (EPA/ARB) 13

2. Inspection and adjustment 15
   2.1 Periodic maintenance schedule 15
   2.2 Periodic inspection and maintenance procedure 17
      2.2.1 Check before starting 17
      2.2.2 inspection after initial 50 hours or one month operation 20
      2.2.3 Inspection every 50 hours or monthly 25
      2.2.4 Inspection every 100 hours six months 30
      2.2.5 Inspection every 150 hours or one year 30
      2.2.6 Inspection every 250 hours or one year 31
      2.2.7 Inspection every 1,000 hours or four years 37
   2.3 Adjusting the no-load maximum or minimum speed 44
   2.4 Sensor Inspection 44
      2.4.1 Oil pressure switch 44
      2.4.2 Thermo switch 44
   2.5 Thermostat inspection 45
   2.6 Adjusting operation 46
      2.6.1 Preliminary precautions 46
      2.6.2 Adjusting operation procedure 46
      2.6.3 Check points and precautions during running 47
   2.7 Long storage 48

3. Troubleshooting 49
   3.1 Preparation before troubleshooting 49
   3.2 Quick reference chart for troubleshooting 50
   3.3 Troubleshooting (Concerning engine and fuel injection equipment) 59

   3.4 Troubleshooting by measuring compression pressure 62
4. Disassembly and reassembly 64
   4.1 Disassembly and reassembly precautions 64
   4.2 Disassembly and reassembly tools 65
      4.2.1 General hand tools... 65
      4.2.2 Special hand tools... 68
      4.2.3 Measuring instruments... 71
      4.2.4 Other material 75
   4.3 Disassembly and reassembly 77
      4.3.1 Disassembly... 77
      4.3.2 Reassembly 89

5. Inspection and servicing of basic engine parts 104
   5.1 Cylinder block 104
      5.1.1 Inspection of parts... 104
      5.1.2 Cleaning of oil holes... 104
      5.1.3 Color check procedure... 104
      5.1.4 Replacement of cap plugs 105
      5.1.5 Cylinder bore measurement... 106
   5.2 Cylinder head 107
      5.2.1 Inspecting the cylinder head 108
      5.2.2 Valve seat correction procedure 109
      5.2.3 Intake/exhaust valves, valve guides 110
      5.2.4 Valve springs 112
      5.2.5 Assembling the cylinder head 113
      5.2.6 Measuring top clearance... 114
      5.2.7 Intake and exhaust rocker arms... 114
      5.2.8 Adjustment of valve clearance 115
   5.3 Piston and piston pins 116
      5.3.1 Piston 116
      5.3.2 Piston pin 117
      5.3.3 Piston rings 118
   5.4 Connecting rod 121
      5.4.1 Inspecting the connection rod 121
      5.4.2 Crank pin metal 122
      5.4.3 Piston pin bushing... 124
      5.4.4 Assembling piston and connecting rod 124
   5.5 Crankshaft and main bearing 125
      5.5.1 Crankshaft... 125
      5.5.2 Main bearing 127

- 5.6 Camshaft and tappets 128
  - 5.6.1 Camshaft.. 128
  - 5.6.2 Tappets 130
- 5.7 Timing gear 131
  - 5.7.1 Inspecting the gears... 131
  - 5.7.2 Gear timing marks.. 131
- 5.8 Flywheel and housing 132
  - 5.8.1 Position of top dead center and fuel injection timing... 132
  - 5.8.2 Damper disc... 133

6. Fuel injection equipment 134
- 6.1 Fuel Injection pump/governor 134
  - 6.1.1 Fuel system diagram.. 134
  - 6.1.2 Fuel injection pump service data and adjustment.. 135
  - 6.1.3 Fuel injection pump structure... 139
  - 6.1.4 Removing a fuel injection pump... 140
  - 6.1.5 Installing a fuel injection pump... 140
  - 6.1.6 Adjusting fuel injection timing... 140
  - 6.1.7 Troubleshooting of fuel injection pump 141
  - 6.1.8 Major faults and troubleshooting.. 141
- 6.2 Fuel feed pump 144
  - 6.2.1 Construction of fuel feed pump 144
  - 6.2.2 Fuel feed pump specifications 144
  - 6.2.3 Disassembly and reassembly of fuel feed pump 145
  - 6.2.4 Fuel feed pump inspection... 145
- 6.3 Fuel filter 147
  - 6.3.1 Fuel filter specifications.. 147
  - 6.3.2 Fuel filter inspection 147
- 6.4 Fuel tank 148

7. Intake and exhaust system 149
- 7.1 Intake system 149
  - 7.1.1 Breather system (A reductor to intake air system of blowby gas)............................ 149
  - 7.1.2 Diaphragm assy inspection.. 150
- 7.2 Exhaust system 151
  - 7.2.1 Construction.. 151
  - 7.2.2 Mixing elbow inspection 151

8. Lubrication system 152
- 8.1 Lubrication system 152
- 8.2 Lube oil pump 153
  - 8.2.1 Lube oil pump construction 153

- 8.2.2 Specifications of lube oil pump 153
- 8.2.3 Lube oil pump disassembly and reassembly 153
- 8.2.4 Lube oil pump inspection 154
- 8.2.5 Pressure control valve construction 154
- 8.3 Lube oil filter 155
  - 8.3.1 Lube oil filter construction 155
  - 8.3.2 Lube oil filter replacement 155
- 8.4 Rotary waste oil pump (Optional) 156

9. Cooling water system 157

- 9.1 Cooling water system 157
- 9.2 Seawater pump 159
  - 9.2.1 Specifications of seawater pump 159
  - 9.2.2 Seawater pump disassembly 159
  - 9.2.3 Seawater pump Inspection 160
  - 9.2.4 Seawater pump reassembly 161
- 9.3 Fresh water pump 162
  - 9.3.1 Fresh water pump construction 162
  - 9.3.2 Specifications of fresh water pump 162
  - 9.3.3 Fresh water pump disassembly 163
  - 9.3.4 WFresh water pump inspection 163
- 9.4 Heat exchanger 165
  - 9.4.1 Heat exchanger construction 165
  - 9.4.2 Disassembly and reassembly of the heat exchanger 165
  - 9.4.3 Heat exchanger inspection 165
- 9.5 Pressure cap and coolant recovery tank 166
  - 9.5.1 Pressure cap construction 166
  - 9.5.2 Pressure cap pressure control 166
  - 9.5.3 Pressure cap inspection 166
  - 9.5.4 Replacing filler neck 167
  - 9.5.5 Function of the coolant recovery tank 168
  - 9.5.6 Specifications of coolant recovery tank 168
  - 9.5.7 Mounting the coolant recovery tank 168
  - 9.5.8 Precautions on usage of the coolant recovery tank 168
- 9.6 Thermostat 169
  - 9.6.1 Functioning of thermostat 169
  - 9.6.2 Thermostat construction 169
  - 9.6.3 Characteristics of thermostat 169
  - 9.6.4 Thermostat inspection 169
  - 9.6.5 Testing the thermostat 169

9.7 Bilge pump and bilge strainer (Optional) 170
    9.7.1 Introduction 170
    9.7.2 Description 171
    9.7.3 Cautions 171
    9.7.4 Assembly procedure 172
    9.7.5 Cautions for assembling 174
    9.7.6 Troubleshooting 175

10. Reduction and reversing gear 176
    10.1 Specifications 176

11. Remote control (Optional) 177
    11.1 Remote control system 177
        11.1.1 Construction of remote control system 177
        11.1.2 Remote control device components 177
    11.2 Remote control installation 179
    11.3 Remote control inspection 181
    11.4 Remote control adjustment 182

12. Electrical system 183
    12.1 Electrical system 183
        12.1.1 Wiring diagram 184
    12.2 Battery 185
    12.3 Starting motor 186
        12.3.1 Specifications 186
        12.3.2 Characteristics 186
        12.3.3 Structure 187
        12.3.4 Wiring diameter of a starting motor 188
    12.4 Alternator standard, 12V/60A 189
        12.4.1 Specifications 189
        12.4.2 Structure 190
        12.4.3 Wiring diagram 191
        12.4.4 Standard output characteristics 191
        12.4.5 Inspection 192
    12.5 Alternator 12V/80A (Optional) 193
        12.5.1 Specifications 193
        12.5.2 Structure 194
        12.5.3 Wiring diagram 195
        12.5.4 Standard output characteristics 195
    12.6 Instrument panel 196
        12.6.1 B-type instrument panel (Optional) 196
    12.7 Warning devices 197

  12.7.1 Oil pressure alarm 197
  12.7.2 Cooling water temperature alarm 198
 12.8 Glow plug 199
 12.9 Electric engine stopping device 200

13. Service standards 201
 13.1 Engine tuning 201
 13.2 Engine body 202
  13.2.1 Cylinder head 202
  13.2.2 Camshaft and gear train 203
  13.2.3 Cylinder block 204
 13.3 Lubricating oil system (Trochoid pump) 207

14. Tightening torque for bolts and nuts 208
 14.1 Main bolt and nut 208
 14.2 Standard bolts and nuts (without lube oil) 208

FOR SAFETY

1. SAFETY LABELS

- Most accidents are caused by negligence of basic safety rules and precautions. For accident prevention, it is important to avoid such causes before development to accidents.
Please read this manual carefully before starting repair or maintenance to fully understand safety precautions and appropriate inspection and maintenance procedures.
Attempting at a repair or maintenance job without sufficient knowledge may cause an unexpected accident.

- It is impossible to cover every possible danger in repair or maintenance in the manual. Sufficient consideration for safety is required in addition to the matters marked  . Especially for safety precautions in a repair or maintenance job not described in this manual, receive instructions from a knowledgeable leader.

- Safety marks used in this manual and their meanings are as follows:

⚠ DANGER   **DANGER**-indicates an imminent hazardous situation which, if not avoided, WILL result in death or serious injury.

⚠ WARNING   **WARNING**-indicates a potentially hazardous situation which, if not avoided, COULD result in death or serious injury.

⚠ CAUTION   **CAUTION**-indicates a potentially hazardous situation which, if not avoided, may result in minor or moderate injury.

- **NOTICE** - indicates that if not observed, the product performance or quality may not be guaranteed.

2. Safety Precautions

(1) SERVICE AREA

- **Sufficient Ventilation**
  Inhalation of exhaust fumes and dust particles may be hazardous to ones health. Running engines welding, sanding, painting, and polishing tasks should be only done in well ventilated areas.

- **Safe / Adequate Work Area**
  The service area should be clean, spacious, level and free from holes in the floor, to prevent "slip" or "trip and fall" type accidents.

- **Clean, orderly arranged place**
  No dust, mud, oil or parts should be left on the floor surface.
  [Failure to Observe]
  An unexpected accident may be caused.

- **Bright, Safely Illuminated Area**
  The work area should be well lit or illuminated in a safe manner. For work in enclosed or dark areas, a "drop cord" should be utilized. The drop cord must have a wire cage to prevent bulb breakage and possible ignition of flammable substances.

- **Safety Equipment**
  Fire extinguisher(s), first aid kit and eye wash / shower station should be close at hand (or easily accessible) in case of an emergency.

(2) WORK - WEAR (GARMENTS)

• Safe Work Clothing

Appropriate safety wear (gloves, special shoes/boots, eye/ear protection, head gear, harness', clothing, etc.) should be used/worn to match the task at hand. Avoid wearing jewelry, unbuttoned cuffs, ties or loose fitting clothes around moving machinery. A serious accident may occur if caught in moving/rotating machinery.

(3) TOOLS

• Appropriate Lifting / Holding

When lifting an engine, use only a lifting device (crane, jack, etc.) with sufficient lifting capacity. Do not overload the device. Use only a chain, cable, or lifting strap as an attaching device. Do not use rope, serious injury may result.

To hold or support an engine, secure the engine to a support stand, test bed or test cart designed to carry the weight of the engine. Do not overload this device, serious injury may result.

Never run an engine without being properly secured to an engine support stand, test bed or test cart, serious injury may result.

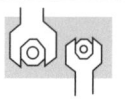

• Appropriate Tools

Always use tools that are designed for the task at hand. Incorrect usage of tools may result in damage to the engine and or serious personal injury.

(4) GENUINE PARTS and MATERIALS

• Genuine Parts

Always use genuine YANMAR parts or YANMAR recommended parts and goods. Damage to the engine, shortened engine life and or personal injury may result.

(5) FASTENER TORQUE

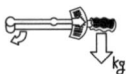

- **Torquing Fasteners**
Always follow the torque values and procedures as designated in the service manual. Incorrect values, procedures and or tools may cause damage to the engine and or personal injury.

(6) Electrical

- **Short Circuits**
Always disconnect the (-) Negative battery cable before working on the electrical system. An accidental "short circuit" may cause damage, fire and or personal injury. Remember to connect the (-) Negative battery cable (back onto the battery) LAST

- **Charging Batteries**
Charging wet celled batteries produces hydrogen gas. Hydrogen gas is extremely explosive. Keep sparks, open flame and any other form of ignition away. Explosion may occur causing severe personal injury.

- **Battery Electrolyte**
Batteries contain sulfuric acid. Do NOT allow it to come in contact with clothing, skin and or eyes, severe burns will result.

(7) WASTE MANAGEMENT

Observe the following instructions with regard to hazardous waste disposal. Negligence of these will have a serious impact on environmental pollution concerns.

1) Waste fluids such as lube oil, fuel and coolant shall be carefully put into separate sealed containers and disposed of properly.
2) Do NOT dispose of waste materials irresponsibly by dumping them into the sewer, overland or into natural waterways.
3) Waste materials such as oil, fuel, coolant, solvents, filter elements and batteries, must be disposed of properly according to local ordinances. Consult the local authorities or reclamation facility.

(8) FURTHER PRECAUTIONS

- **Fueling / Refueling**
  Keep sparks, open flames or any other form of ignition (match, cigarette, etc.) away when fueling/refueling the unit. Fire and or an explosion may result.

- **Hot Surfaces.**
  Do NOT touch the engine (or any of its components) during running or shortly after shutting it down. Scalding / serious burns may result. Allow the engine to cool down before attempting to approach the unit.

- **Rotating Parts**
  Be careful around moving/rotating parts. Loose clothing, jewelry, ties or tools may become entangled causing damage to the engine and or severe personal injury.

- **Preventing burns from scalding**
  1) Never open the filler cap shortly after shutting the engine down. Steam and hot water will spurt out and seriously burn you. Allow the engine to cool down before attempt to open the filler cap.
  2) Securely tighten the filler cap after checking the cooling water. Steam can spurt out during engine running, if tightening loose.

- **Safety Label Check**
  Pay attention to the product safety label.
  A safety label (caution plate) is affixed on the product for calling special attention to safety.
  If it is missing or illegible, always affix a new one.

3. Precautions for Service Work

(1) Precautions for Safety
Read the safety precautions given at the beginning of this manual carefully and always mind safety in work.

(2) Preparation for Service Work
Preparation is necessary for accurate, efficient service work. Check the customer ledger file for the history of the engine.
- Preceding service date
- Period/operation hours after preceding service
- Problems and actions in preceding service
- Replacement parts expected to be required for service
- Recording form/check sheet required for service

(3) Preparation before Disassembly
- Prepare general tools, special service tools, measuring instruments, oil, grease, non-reusable parts, and parts expected to be required for replacement.
- When disassembling complicated portions, put match-marks and other marks at places not adversely affecting the function for easy reassembly.

(4) Precautions in Disassembly
- Each time a parts is removed, check the part installed state, deformation, damage, roughening, surface defect, etc.
- Arrange the removed parts orderly with clear distinction between those to be replaced and those to be used again.
- Parts to be used again shall be washed and cleaned sufficiently.
- Select especially clean locations and use clean tools for disassembly of hydraulic units such as the fuel injection pump.

(5) Precautions for Inspection and Measurement
Inspect and measure parts to be used again as required to determine whether they are reusable or not.

(6) Precautions for Reassembly
- Reassemble correct parts in correct order according to the specified standards (tightening torques, and adjustment standards). Apply oil important bolts and nuts before tightening when specified.
- Always use genuine parts for replacement.
- Always use new oil seals, O-rings, packing and cotter pins.
- Apply sealant to packing depending on the place where they are used. Apply of grease to sliding contact portions, and apply grease to oil seal lips.

(7) Precautions for Adjustment and Check
Use measuring instruments for adjustment to the specified service standards.

1. General

1.1 Exterior views

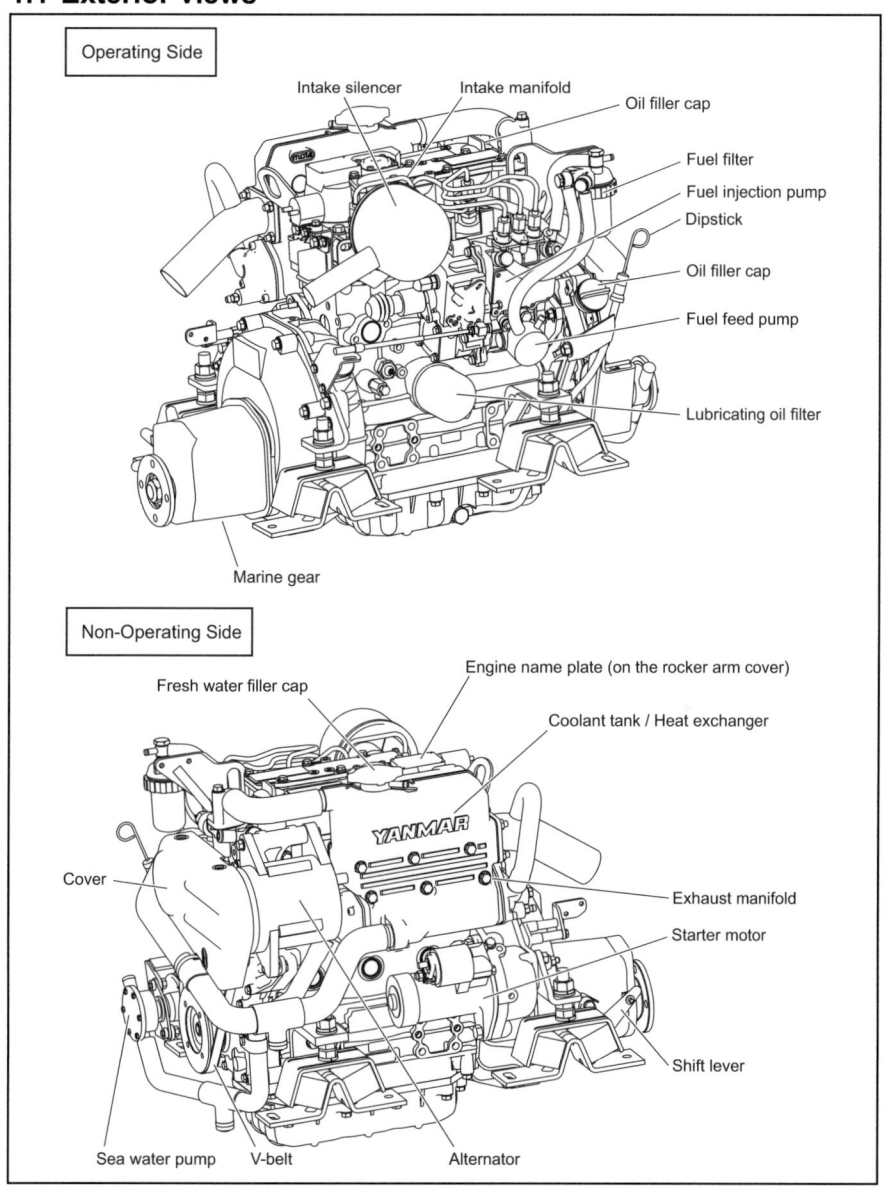

<Note> This illustration shows the 3YM30 with Yanmar marine gear (Model:KM2P-1).

1. General

1.2 Specifications

Official engine model name	unit	3YM30		3YM20	
Company internal model name	-	3YM30	3YM30C	3YM20	3YM20C
Marine gear model	-	KM2P-1	SD20	KM2P-1	SD20
Use	-	Pleasure use			
Type	-	Vertical water cooled 4 cycle diesel engine			
Combustion system	-	Indirect injection			
Air charging	-	Naturally aspirated			
Number of cylinders	-	3			
Bore x stroke	mm(inch)	76 x 82 (2.99 x 3.23)		70 x 74 (2.76 x 2.91)	
Displacement	L	1.115		0.854	
Continuous power — Output at crankshaft / Engine speed	kW(HP)/ min^{-1}	20.1(27.3) / 3489 (at Fuel temp. 25°C) *		14.7(20.0) / 3489 (at Fuel temp. 25°C) *	
Fuel stop power — Output at crankshaft / Engine speed	kW(HP)/ min^{-1}	22.1(30) / 3600 (at Fuel temp. 25°C) * 21.3(29.0) / 3600 (at Fuel temp. 40°C) **		16.2(22) / 3600 (at Fuel temp. 25°C) * 15.3(20.8) / 3600 (at Fuel temp. 40°C) **	
Fuel stop power — Output at propeller shaft / Engine speed	kW(HP)/ min^{-1}	21.4(29.1) / 3600 (at Fuel temp. 25°C) * 20.7(28.1) / 3600 (at Fuel temp. 40°C) **		15.7(21.3) / 3600 (at Fuel temp. 25°C) * 14.9(20.2) / 3600 (at Fuel temp. 40°C) **	
Installation	-	Flexible mounting			
Fuel injection timing	deg b.T.D.C.	FID 16±1 (FIC-Air : 18±1)		FID 22±1 (FIC-Air : 24±1)	
Fuel injection opening pressure	MPa (kgf/cm^2)	11.8$^{+0.98}/_{-0}$ (120$^{+10}/_{-0}$)		12.3$^{+0.98}/_{-0}$ (125$^{+10}/_{-0}$)	
Main power take off	-	At Flywheel side			
Direction of rotation — Crankshaft	-	Counter-clockwise viewed from stern			
Direction of rotation — Propeller shaft (Ahead)	-	Clockwise viewed from stern			
Cooling system	-	Fresh water cooling with heat exchanger			
Lubrication system	-	Complete enclosed forced lubrication			
Cooling water capacity (fresh water)	L(quart)	Engine:4.9 (5.2), Coolant recovery tank : 0.8 (0.8)		Engine:4.1 (4.3), Coolant recovery tank : 0.8 (0.8)	
Lubricating oil capacity (engine) — Rake angle	deg.	at rake angle 8 deg.	at rake angle 0 deg.	at rake angle 8 deg.	at rake angle 0 deg.
Lubricating oil capacity (engine) — Total (Note 4)	L(quart)	2.8 $^0/_{-0.2}$ (3.0 $^0/_{-0.2}$)	2.5 $^0/_{-0.2}$ (2.6 $^0/_{-0.2}$)	2.7 $^0/_{-0.2}$ (2.9 $^0/_{-0.2}$)	2.4 $^0/_{-0.2}$ (2.5 $^0/_{-0.2}$)
Lubricating oil capacity (engine) — Effective (Note 5)		1.4 (1.5)	1.5 (1.6)	1.4 (1.5)	1.5 (1.6)
Starting system — Type	-	Electric			
Starting system — Starting motor	V-kW	DC 12V-1.4 kW			
Starting system — AC generator	V-A	12V-60A (12V-80A optional)			
Engine Dimension — Overall length	mm(inch)	715 (28.1)	715 (28.1)	698 (27.5)	698 (27.5)
Engine Dimension — Overall width	mm(inch)	463 (18.2)	463 (18.2)	463 (18.2)	463 (18.2)
Engine Dimension — Overall height	mm(inch)	545 (21.5)	545 (21.5)	528 (20.8)	528 (20.8)
Engine dry mass (include marine gear)	kg	133	157 (with SD20)	120	144 (with SD20)

1. General

Official engine model name	unit	2YM15	
Company internal model name	-	2YM15	2YM15C
Marine gear model	-	KM2P-1	SD20
Use	-	Pleasure use	
Type	-	Vertical water cooled 4 cycle diesel engine	
Combustion system	-	Indirect injection	
Air charging	-	Naturally aspirated	
Number of cylinders	-	2	
Bore x stroke	mm(inch)	70 x 74 (2.76 x 2.91)	
Displacement	L	0.570	
Continuous power — Output at crankshaft / Engine speed	kW(HP)/ min^{-1}	9.4(12.8) / 3489 (at Fuel temp. 25°C) *	
Fuel stop power — Output at crankshaft / Engine speed	kW(HP)/ min^{-1}	10.3(14.0) / 3600 (at Fuel temp. 25°C) * 10.0(13.6) / 3600 (at Fuel temp. 40°C) **	
Fuel stop power — Output at propeller shaft / Engine speed	kW(HP)/ min^{-1}	10.0(13.6) / 3600 (at Fuel temp. 25°C) * 9.7(13.2) / 3600 (at Fuel temp. 40°C) **	
Installation	-	Flexible mounting	
Fuel injection timing	deg b.T.D.C.	FID 21±1 (FIC-Air : 23±1)	
Fuel injection opening pressure	MPa (kgf/cm^2)	12.3$^{+0.98}/_{-0}$ (125$^{+10}/_{-0}$)	
Main power take off	-	At Flywheel side	
Direction of rotation — Crankshaft	-	Counter-clockwise viewed from stern	
Direction of rotation — Propeller shaft (Ahead)	-	Clockwise viewed from stern	
Cooling system	-	Fresh water cooling with heat exchanger	
Lubrication system	-	Complete enclosed forced lubrication	
Cooling water capacity (fresh water)	L(quart)	Engine:3.0 (3.2), Coolant recovery tank : 0.8 (0.8)	
Lubricating oil capacity (engine) — Rake angle	deg.	at rake angle 8 deg.	at rake angle 0 deg.
Lubricating oil capacity (engine) — Total (Note 4)	L(quart)	2.0 $^0/_{-0.2}$ (2.1 $^0/_{-0.2}$)	1.8 $^0/_{-0.2}$ (1.9 $^0/_{-0.2}$)
Lubricating oil capacity (engine) — Effective (Note 5)		0.95 (1.0)	1.5 (1.6)
Starting system — Type	-	Electric	
Starting system — Starting motor	V-kW	DC 12V-1.4 kW	
Starting system — AC generator	V-A	12V-60A (12V-80A optional)	
Engine Dimension — Overall length	mm(inch)	613 (24.1)	613 (24.1)
Engine Dimension — Overall width	mm(inch)	463 (18.2)	463 (18.2)
Engine Dimension — Overall height	mm(inch)	528 (20.8)	528 (20.8)
Engine dry mass (include marine gear)	kg	115	134 (with SD20)

2-2R1

(Note)
1. Rating condition : ISO 3046-1, 8665
2. 1HP (metric horse power) ≒ 0.7355 kW
3. Fuel condition : Density at 15°C = 0.842
    * Fuel temperature 25°C at the inlet of the fuel injection pump. (ISO 3046-1)
    ** Fuel temperature 40°C at the inlet of the fuel injection pump. (ISO 8665)
4. The "Total" oil quantity includes: oil in oil pan and oil in channels, coolers and filter.
5. The effective amount of oil shows the difference in maximum scale of the dipstick and minimum scale.

1. General

1.3 Fuel oil, lubricating oil and cooling water

1.3.1 Fuel oil

IMPORTANT:
Only use the recommended fuel to obtain the best engine performance and prevent damage of parts, also prevent air pollution.

(1) Selection of fuel oil
Use the following diesel fuels for best engine performance:
BS 2869 A1 or A2

Fuels equivalent to Japanese Industrial Standard, JIS. No. K2204-2

Fuel cetane number should be 45 or greater

(2) Fuel handling
- Water and dust in the fuel oil can cause operation failure. Use containers which are clean inside to store fuel oil. Store the containers away from rain water and dust.
- Before supplying fuel, let the fuel container rest for several hours so that water and dust in the fuel are deposited on the bottom. Pump up only the clean fuel.

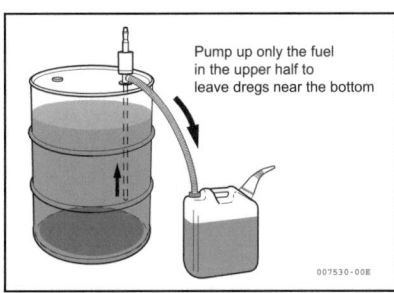

Pump up only the fuel in the upper half to leave dregs near the bottom

(3) Fuel tank
Fuel tank inside should be always clean enough and dry it inside for the first use.
Drain the water according to the maintenance schedule with a drain cock.

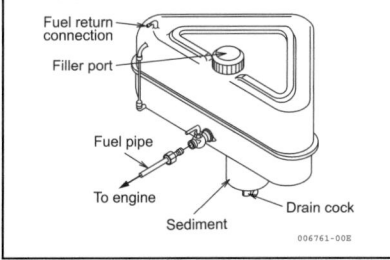

1.3.2 Lubricating oil

IMPORTANT:
Use of other than the specified engine oil may cause inner parts seizure or early wear, leading to shorten the engine service life.

(1) Selection of engine lube oil
Use the following engine oil
- API classification .... CD or better
  (Standards of America Petroleum Institute)
- SAE viscosity .... 10W-30 or 15W-40
  (Standard of Society of Automotive Engineering)

Engine oil with 10W-30 or 15W-40 can be used throughout the year.
(Refer to the right figure.)

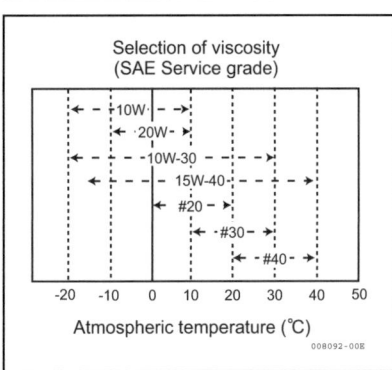

Selection of viscosity (SAE Service grade)

Atmospheric temperature (°C)

1. General

(2) Selection of marine gear lube oil
Use the following engine oil
- API classification .... CD or better
  (Standards of America Petroleum Institute)
- SAE viscosity .... #20 or #30
  (Standard of Society of Automotive Engineering)

(3) Selection of lube oil for sail drive unit
API service grade GL4, 5
SAE viscosity .... #90 or 80W-90
or QuickSilver High Performance Gear Lube
QuickSilver® is registered trademark of Brunswick Corporation.

(4) Handling of engine oil
- Carefully store and handle the oil so as to prevent dust or dirt entrance. When supplying the oil, pay attention and clean around the filler port.
- Do not mix different types of oil as it may adversely affect the lubricating performance.

When touching engine oil by hand, the skin of the hand may become rough. Be careful not to touch oil with your hands without protective gloves. If touch, wash your hands with soap and water thoroughly.

1.3.3 Cooling water

Use clean soft water and be sure to add the Long Life Coolant Antifreeze (LLC) in order to prevent rust built up and freezing. If there is any doubt over the water quality, distilled water or pre-mixed coolant should be used.
The coolants / antifreezes, which are good performance for example, are shown below.
- TEXACO LONG LIFE COOLANT ANTIFREEZE, both standard and pre-mixed.
  Product codes 7997 and 7998
- HAVOLINE EXTENDED LIFE ANTIFREEZE / COOLANT
  Product code 7994

IMPORTANT:
- Be sure to add Long Life Coolant Antifreeze (LLC) to soft water. In cold season, the LLC is especially important. Without LLC, cooling performance will decrease due to scale and rust in the cooling water line. Without LLC, cooling water will freeze and expand to break the cooling line.
- Be sure to use the mixing ratios specified by the LLC manufacturer for your temperature range.
- Do not mix different types (brand) of LLC, chemical reactions may make the LLC useless and engine trouble could result.
- Replace the cooling water every once a year.

When handling Long Life Coolant Antifreeze, wear protective rubber gloves not to touch it. If LLC gets eyes or skin, wash with clean water at once.

1.4 Engine outline

(1) 3YM30 (with KM2P-1 marine gear)

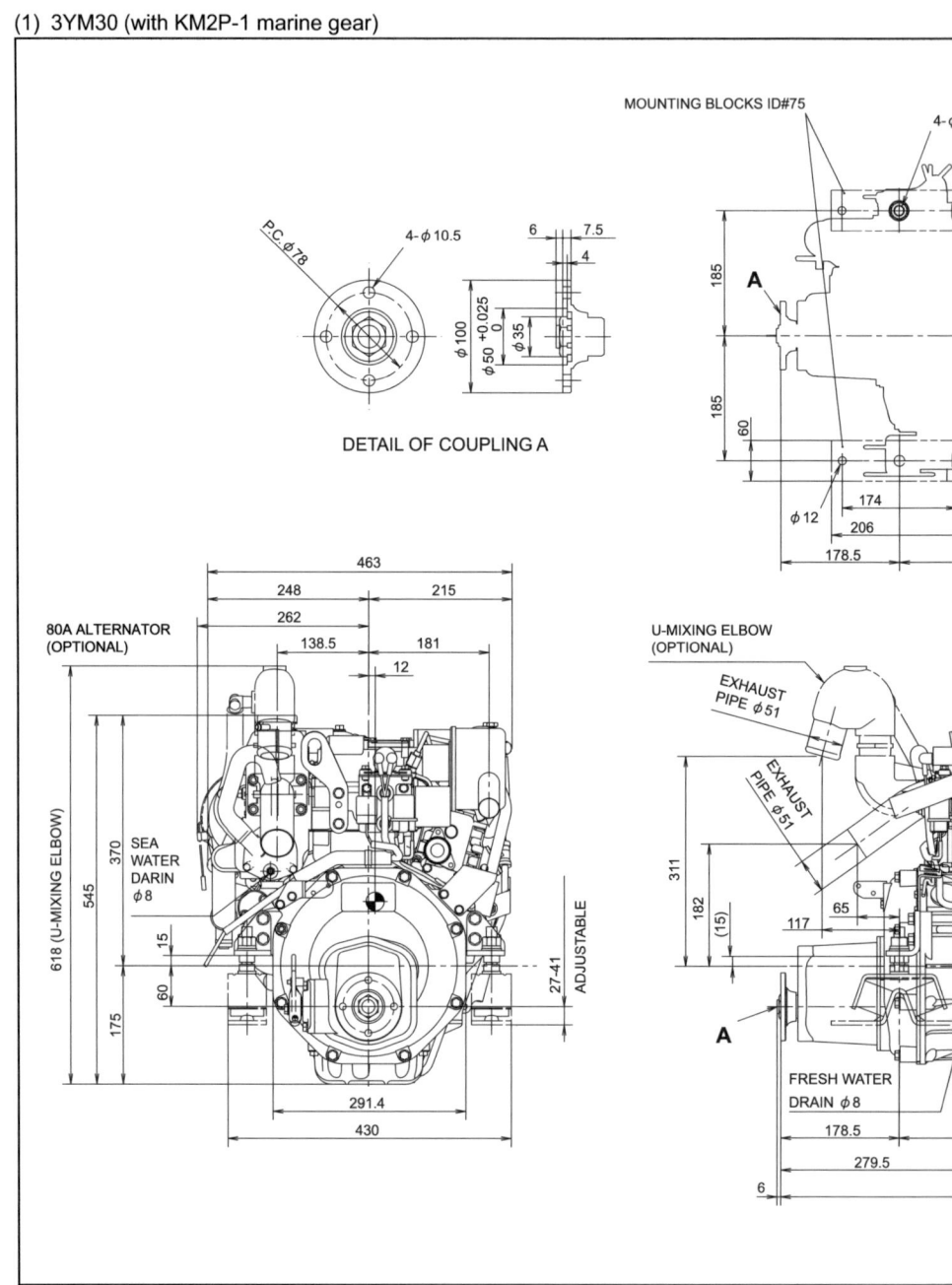

1. General

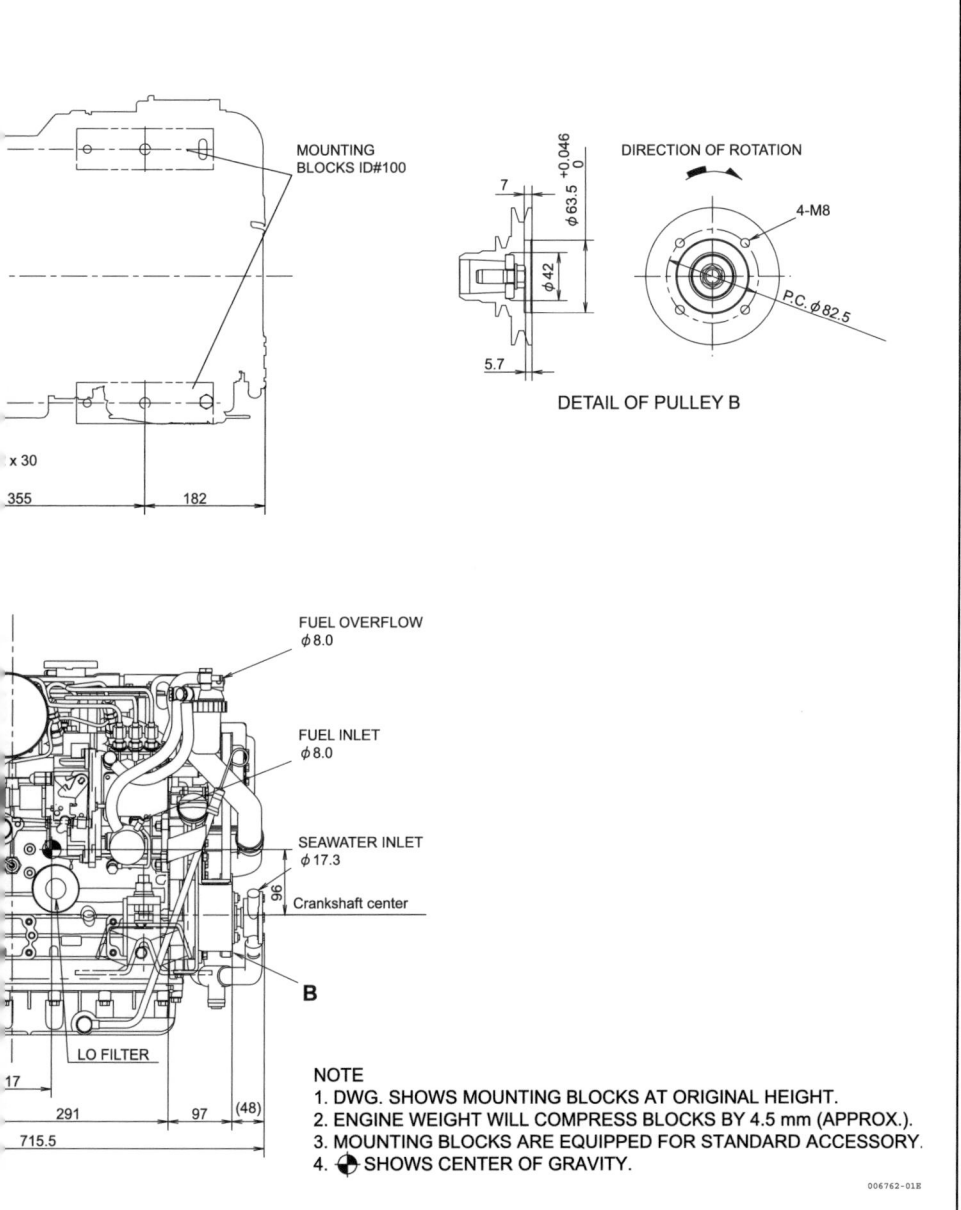

DETAIL OF PULLEY B

NOTE
1. DWG. SHOWS MOUNTING BLOCKS AT ORIGINAL HEIGHT.
2. ENGINE WEIGHT WILL COMPRESS BLOCKS BY 4.5 mm (APPROX.).
3. MOUNTING BLOCKS ARE EQUIPPED FOR STANDARD ACCESSORY.
4. ⊕ SHOWS CENTER OF GRAVITY.

(2) 3YM30C (with SD20 sail drive)

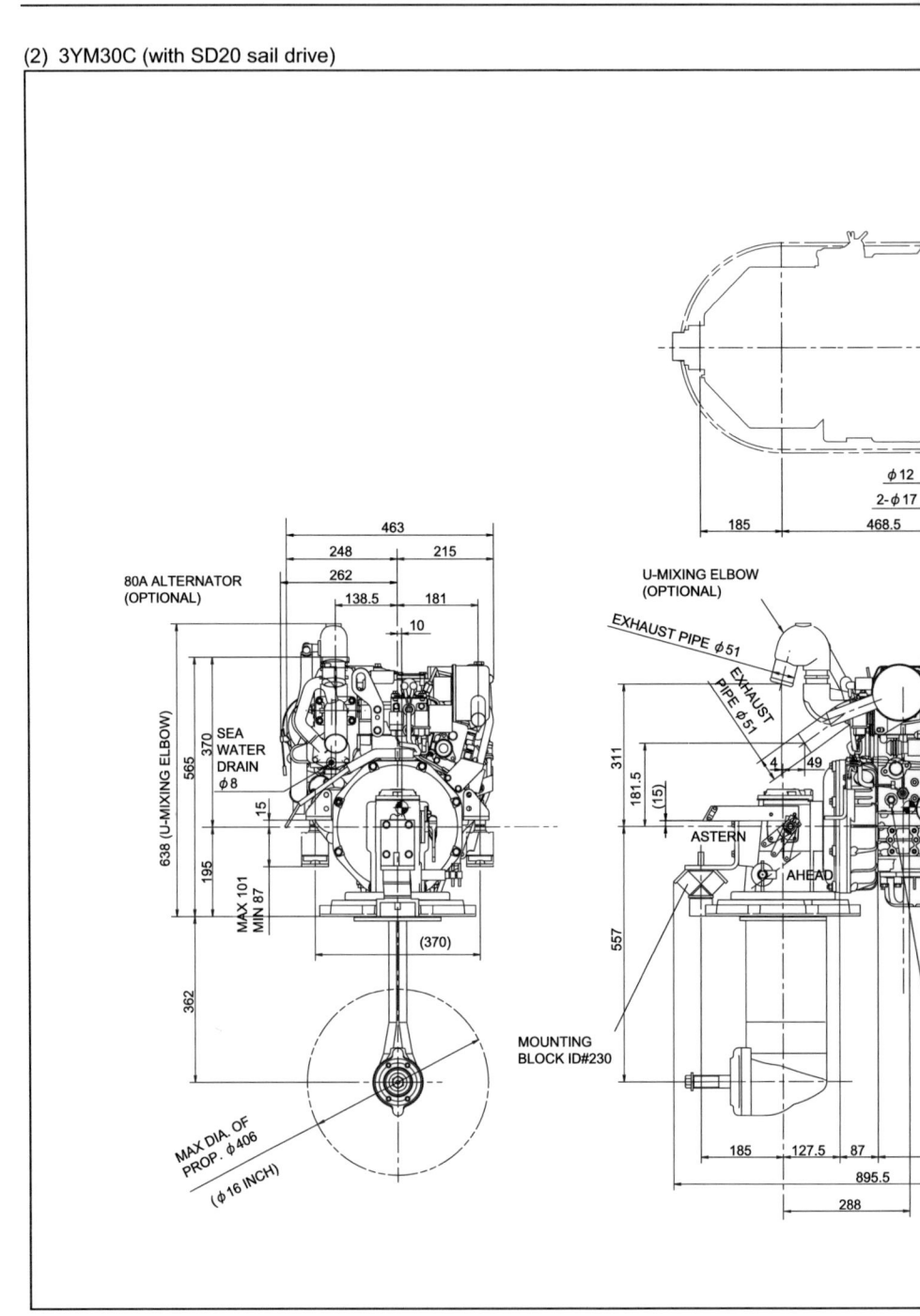

1. General

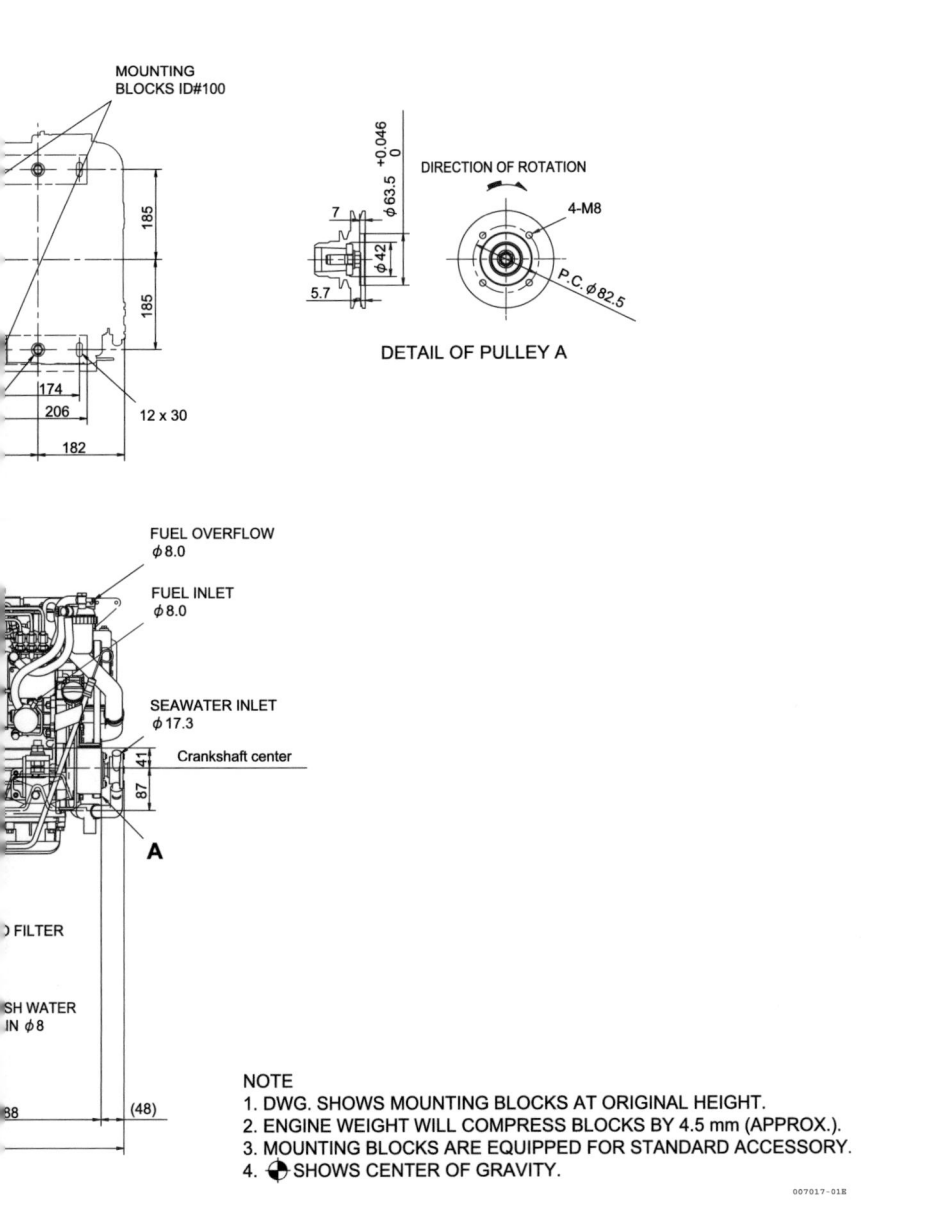

NOTE
1. DWG. SHOWS MOUNTING BLOCKS AT ORIGINAL HEIGHT.
2. ENGINE WEIGHT WILL COMPRESS BLOCKS BY 4.5 mm (APPROX.).
3. MOUNTING BLOCKS ARE EQUIPPED FOR STANDARD ACCESSORY.
4. ⊕ SHOWS CENTER OF GRAVITY.

(3) 3YM20 (with KM2P-1 marine gear)

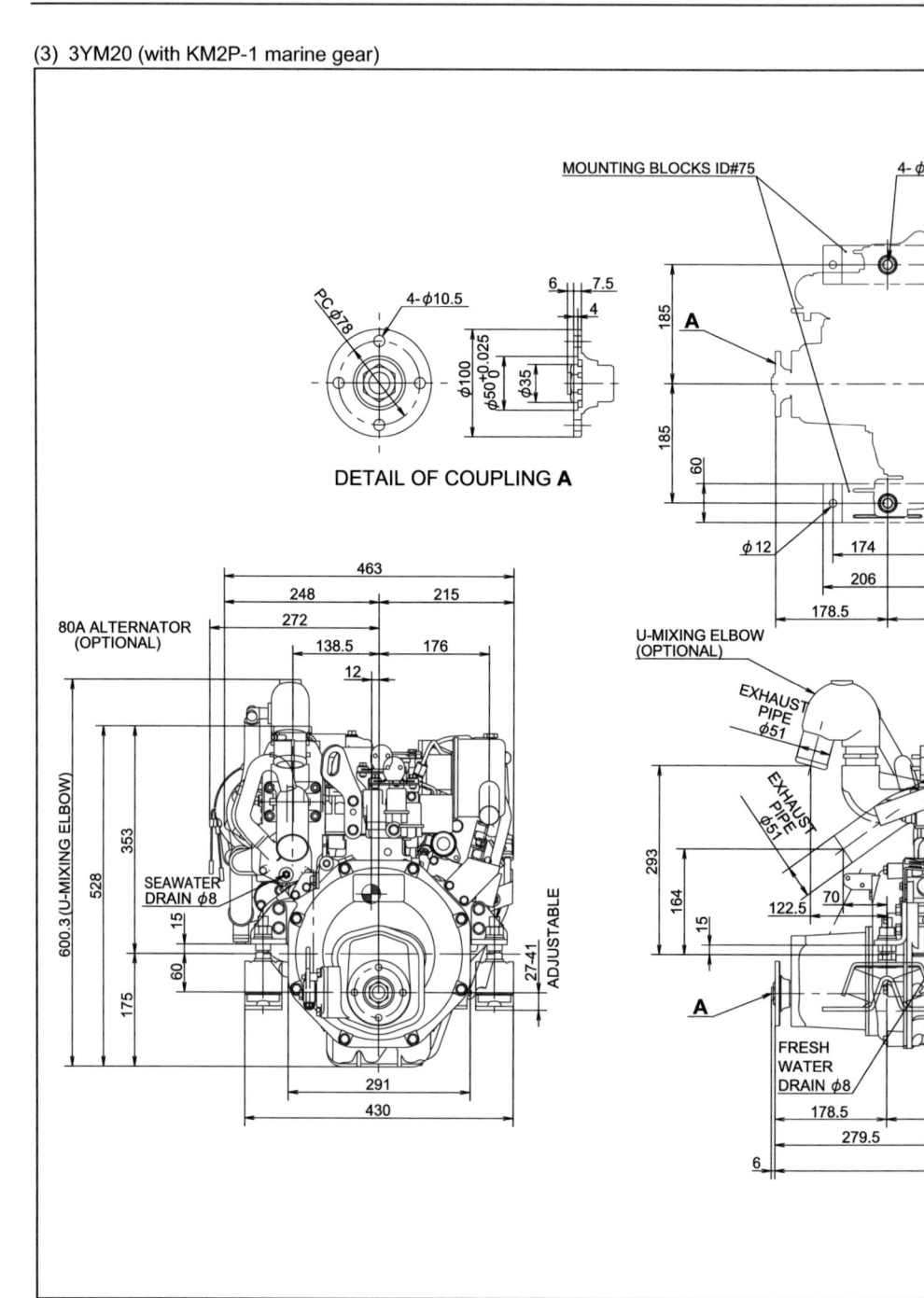

1. General

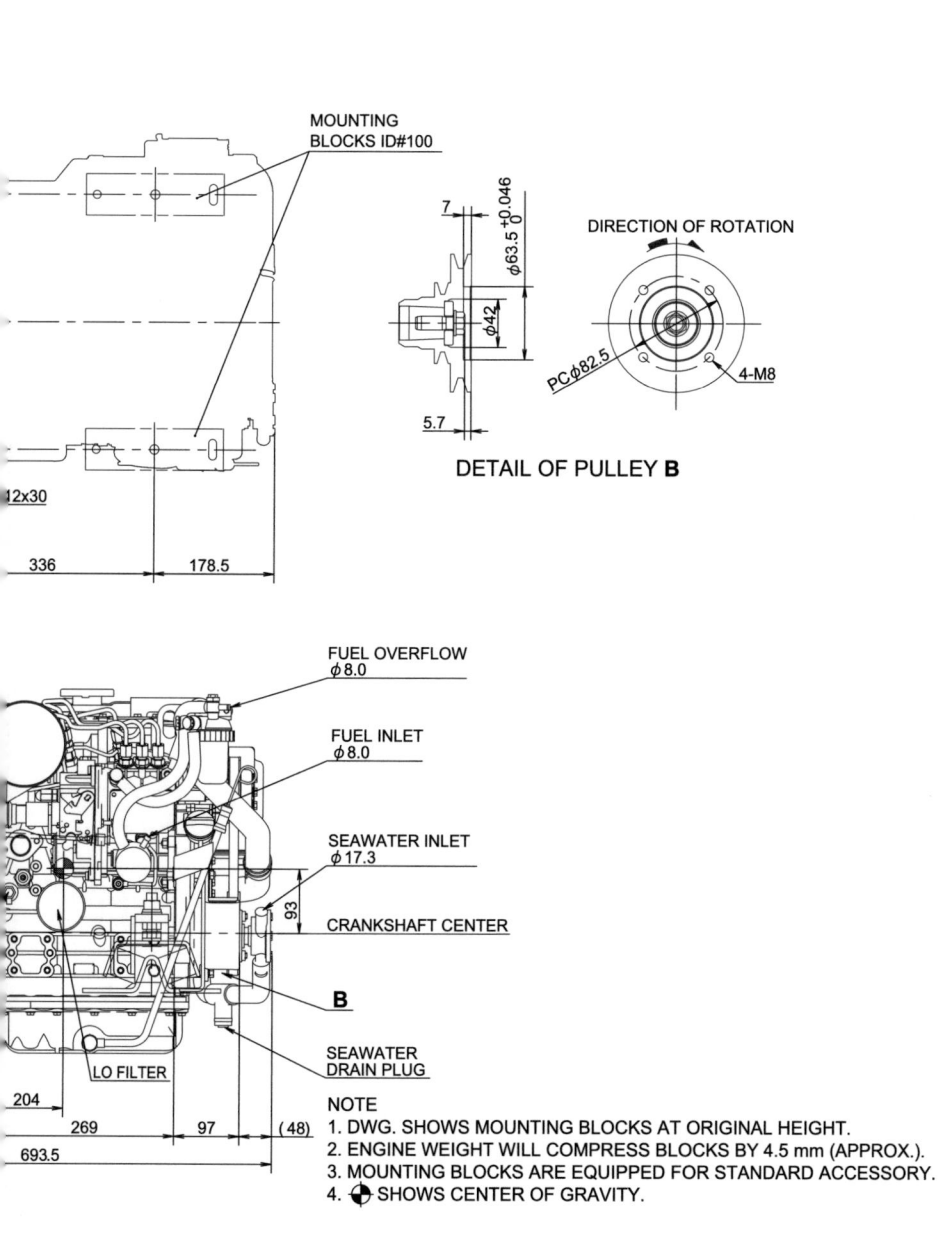

DETAIL OF PULLEY B

NOTE
1. DWG. SHOWS MOUNTING BLOCKS AT ORIGINAL HEIGHT.
2. ENGINE WEIGHT WILL COMPRESS BLOCKS BY 4.5 mm (APPROX.).
3. MOUNTING BLOCKS ARE EQUIPPED FOR STANDARD ACCESSORY.
4. ⊕ SHOWS CENTER OF GRAVITY.

(4) 3YM20C (with SD20 sail drive)

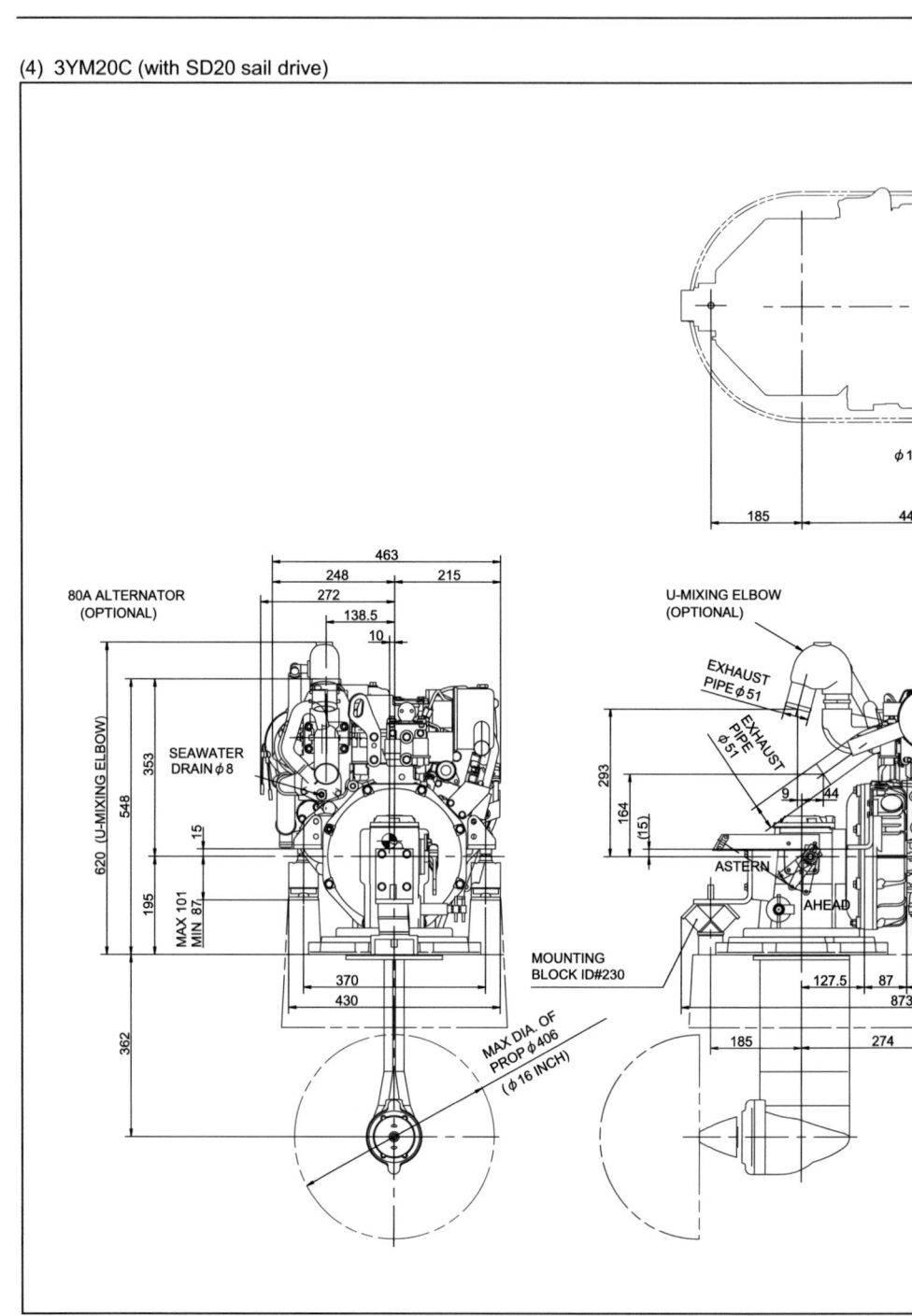

1. General

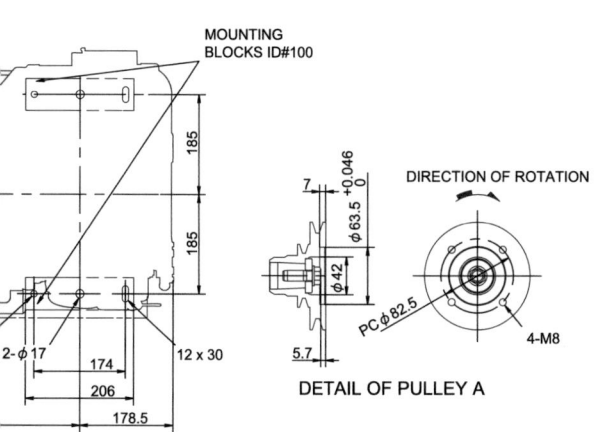

DETAIL OF PULLEY A

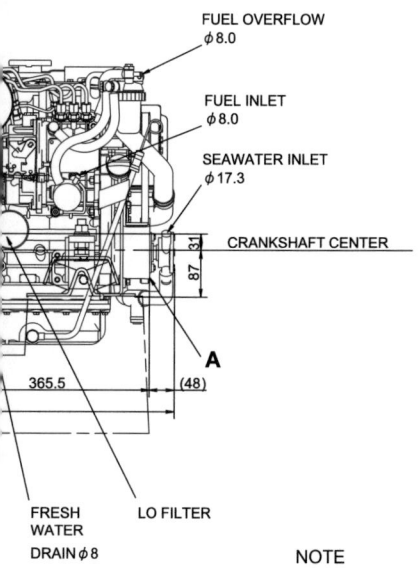

FUEL OVERFLOW
φ8.0

FUEL INLET
φ8.0

SEAWATER INLET
φ17.3

CRANKSHAFT CENTER

FRESH WATER DRAIN φ8

LO FILTER

NOTE
1. DWG. SHOWS MOUNTING BLOCKS AT ORIGINAL HEIGHT.
2. ENGINE WEIGHT WILL COMPRESS BLOCKS BY 4.5 mm (APPROX.).
3. MOUNTING BLOCKS ARE EQUIPPED FOR STANDARD ACCESSORY.
4. ⬥ SHOWS CENTER OF GRAVITY.

(5) 2YM15 (with KM2P-1 marine gear)

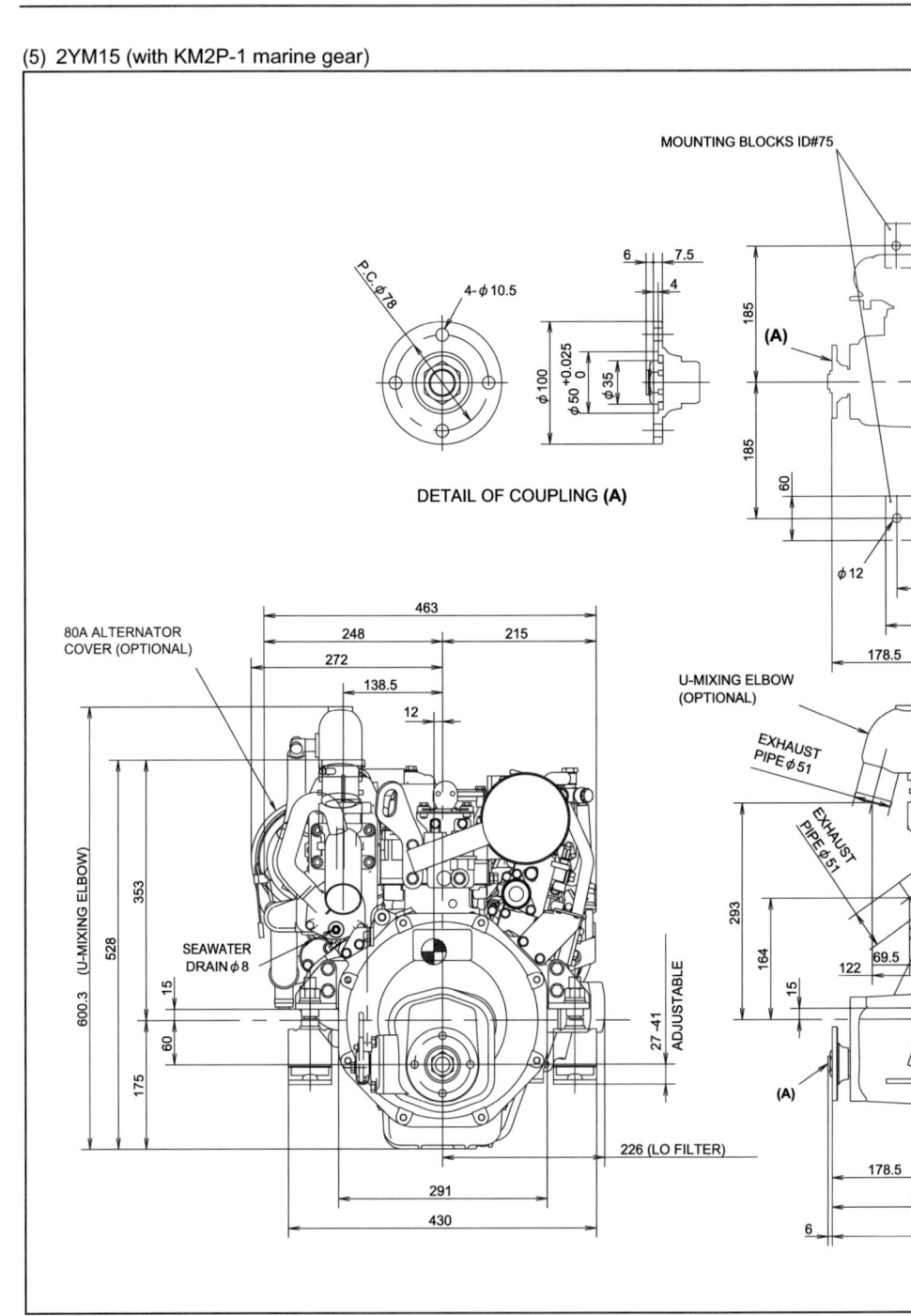

1. General

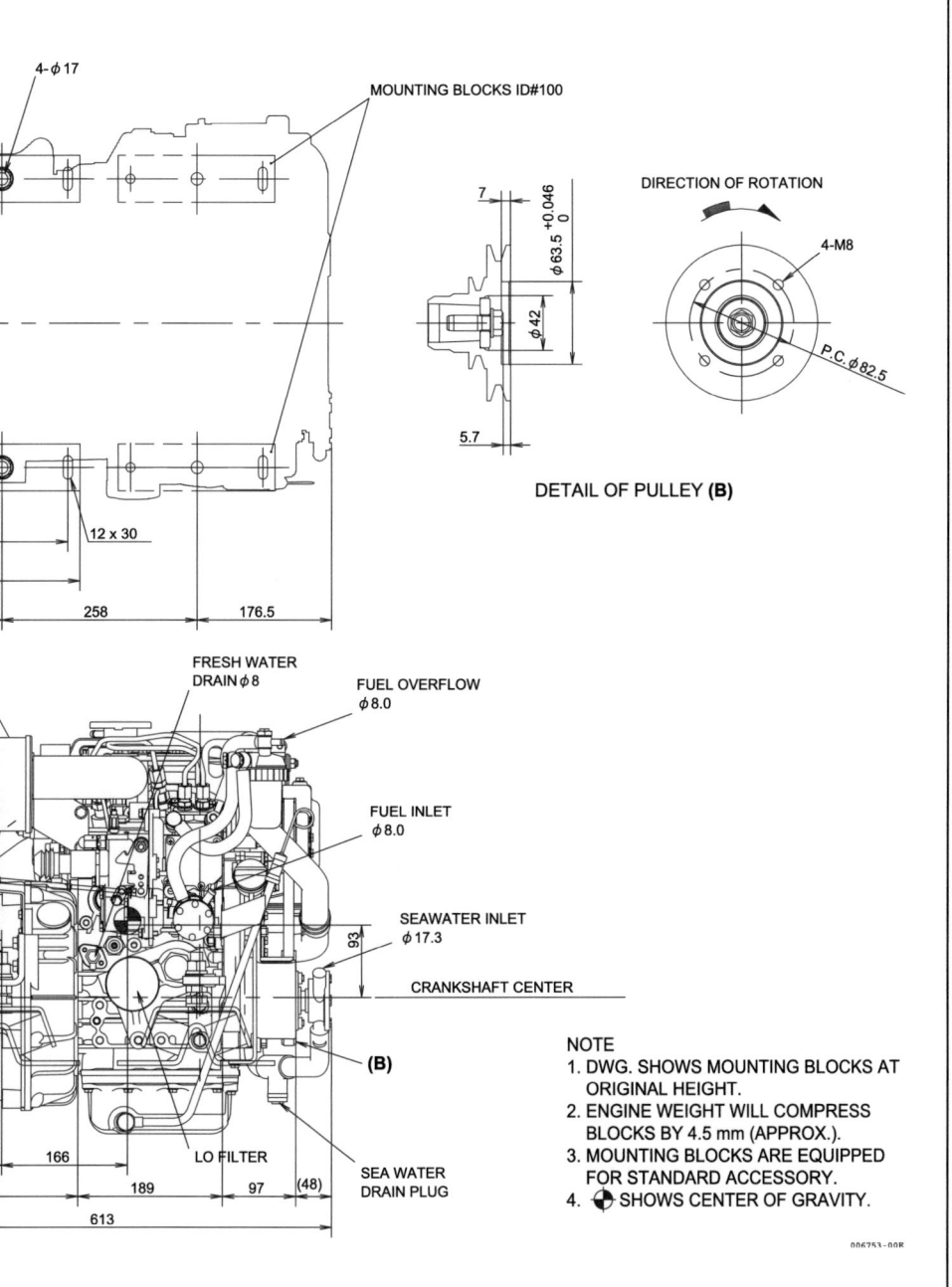

(6) 2YM15C (with SD20 sail drive)

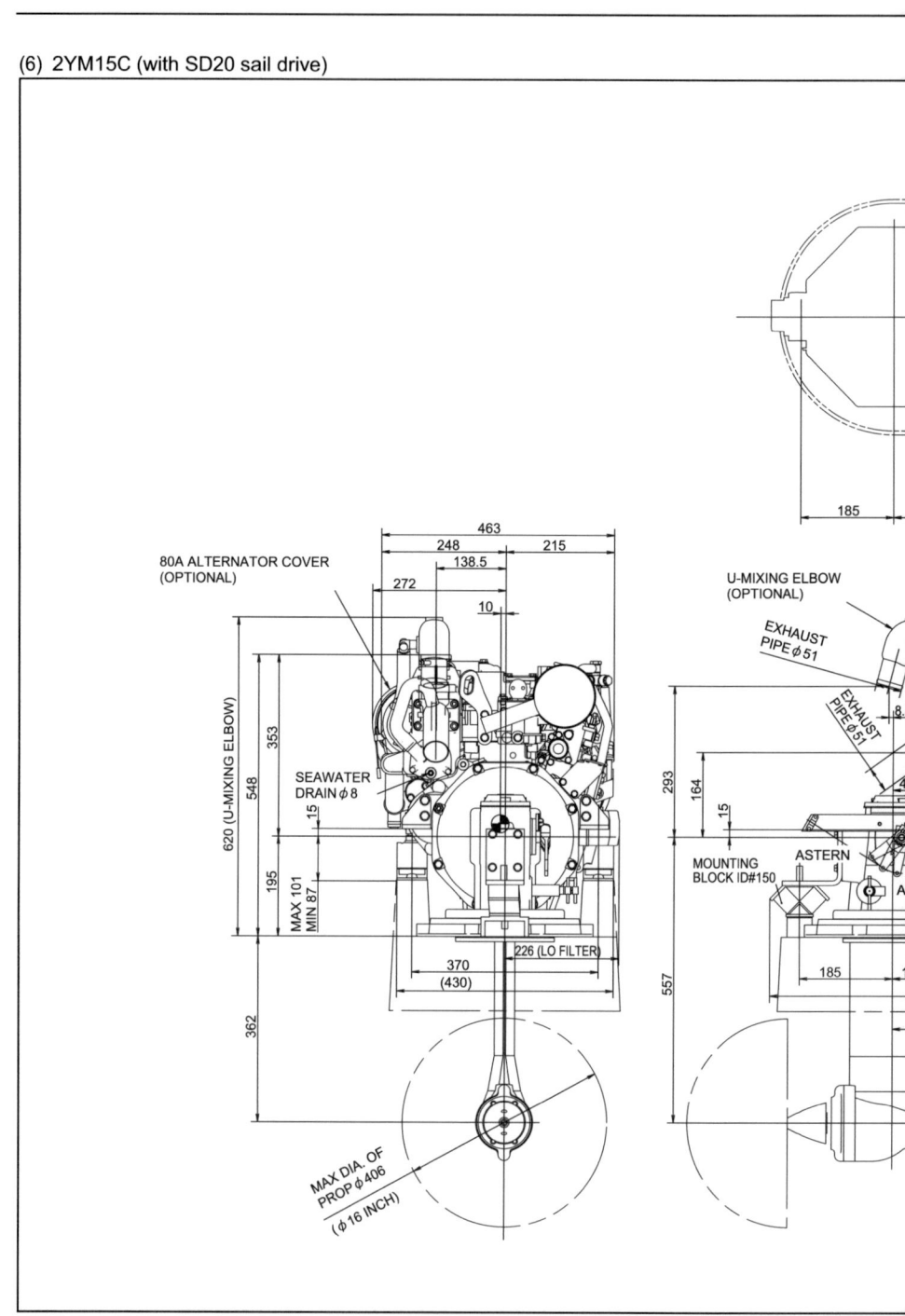

1. General

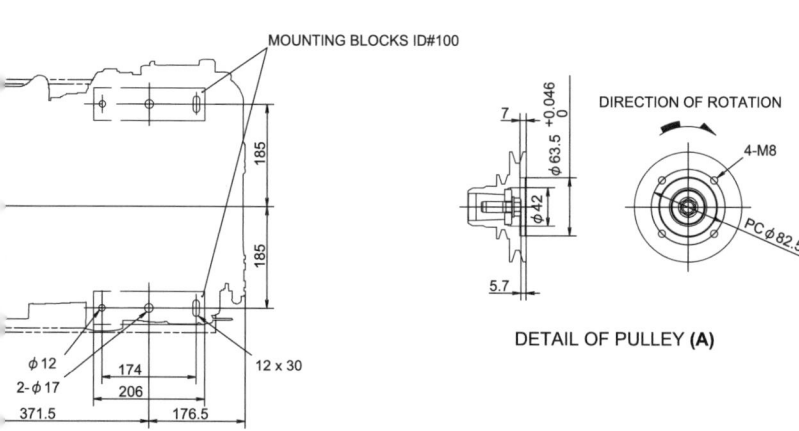

DETAIL OF PULLEY (A)

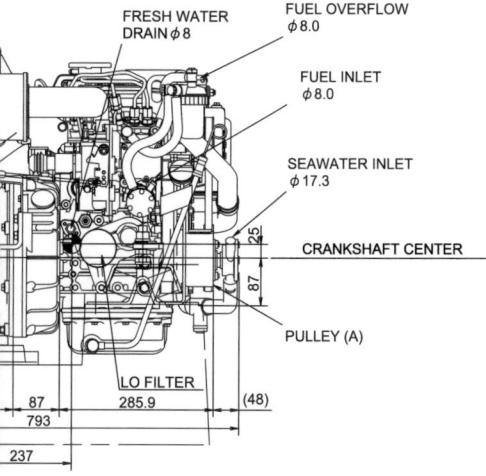

NOTE
1. DWG. SHOWS MOUNTING BLOCKS AT ORIGINAL HEIGHT.
2. ENGINE WEIGHT WILL COMPRESS BLOCKS BY 4.5 mm (APPROX.).
3. MOUNTING BLOCKS ARE EQUIPPED FOR STANDARD ACCESSORY.
4. ⊕ SHOWS CENTER OF GRAVITY.

1.5 Piping diagrams

(1) 3YM30/3YM20 (with KM2P-1 marine gear)

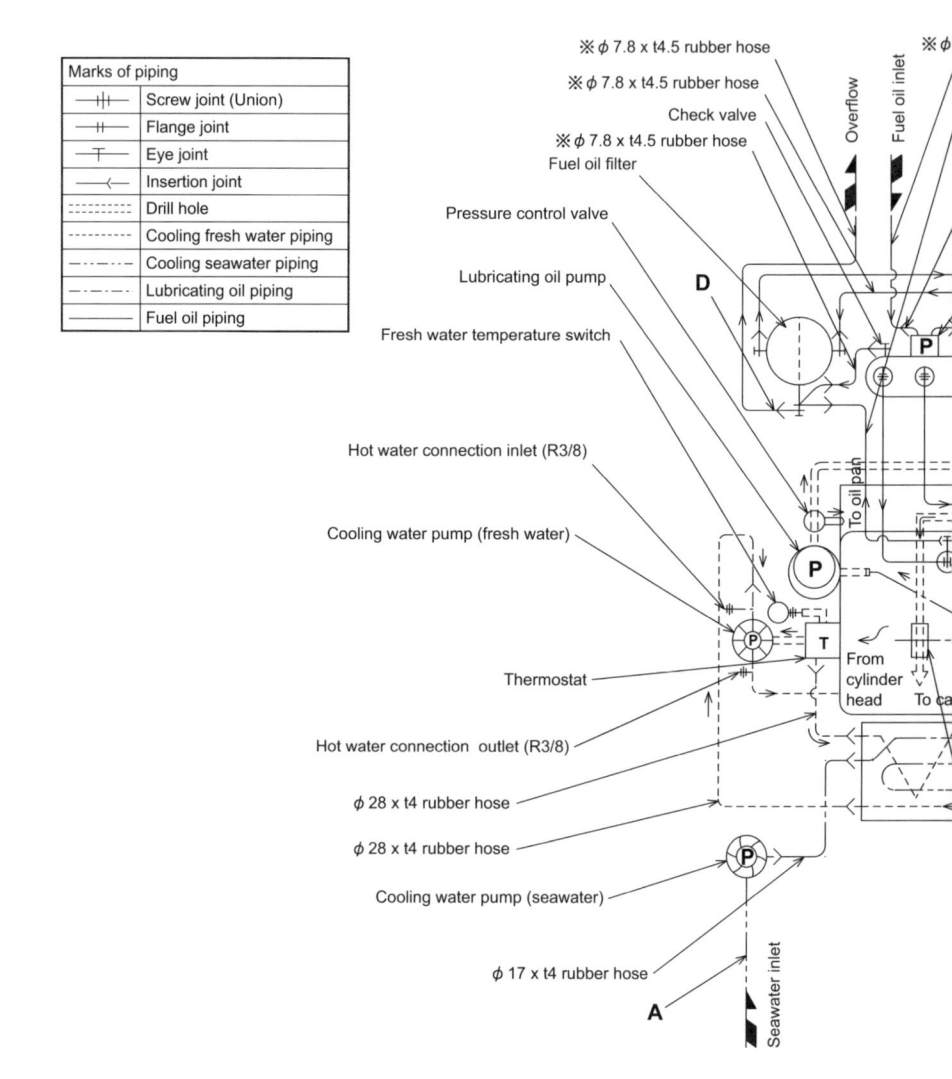

Notes
1. Dimension of steel pipe : outer dia. x thickness.
   Dimension of rubber pipe : inner dia. x thickness.
2. Fuel rubber pipes (marked ※) satisfy EN/ISO7840.

1. General

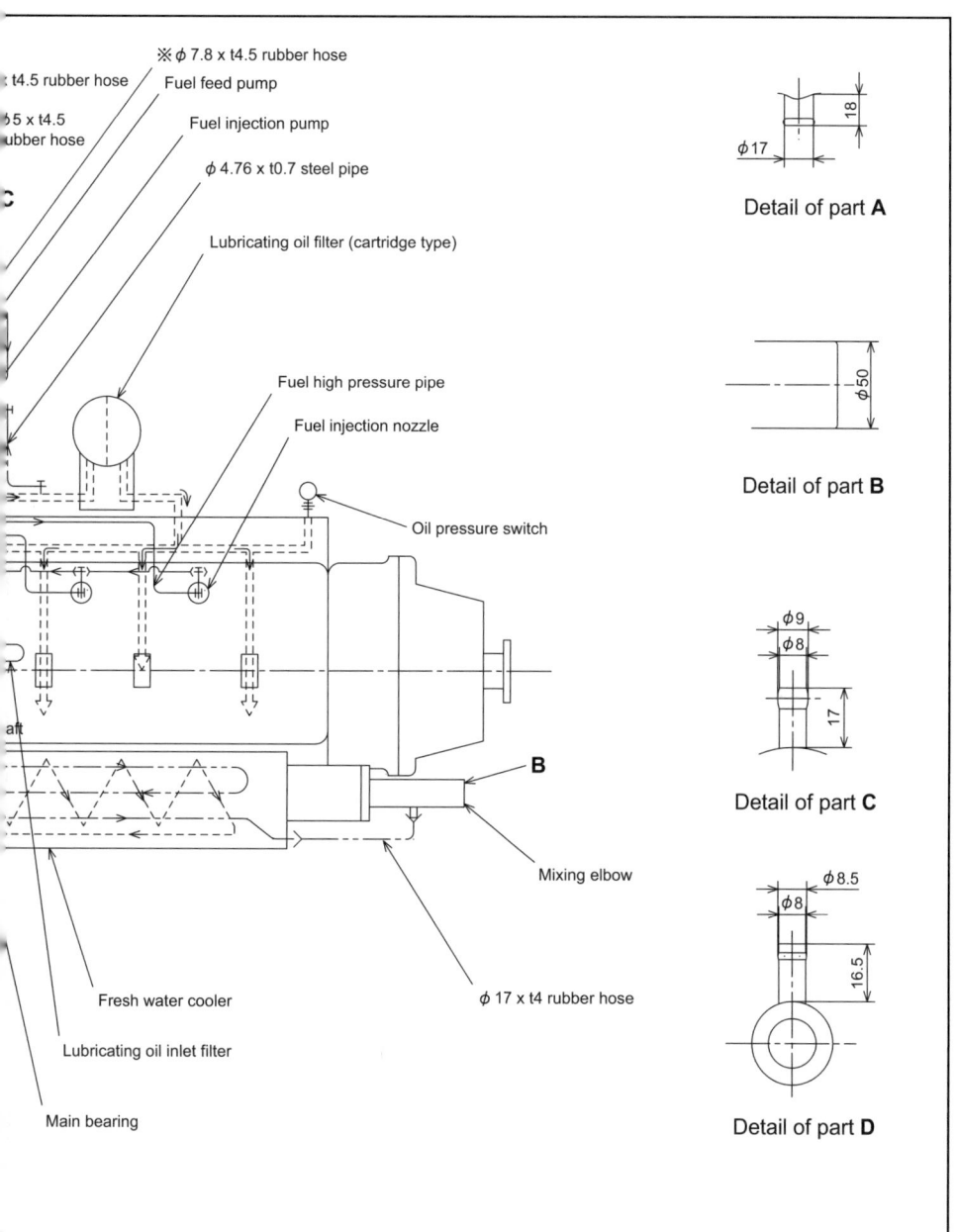

(2) 3YM30C/3YM20C (with sale drive SD20)

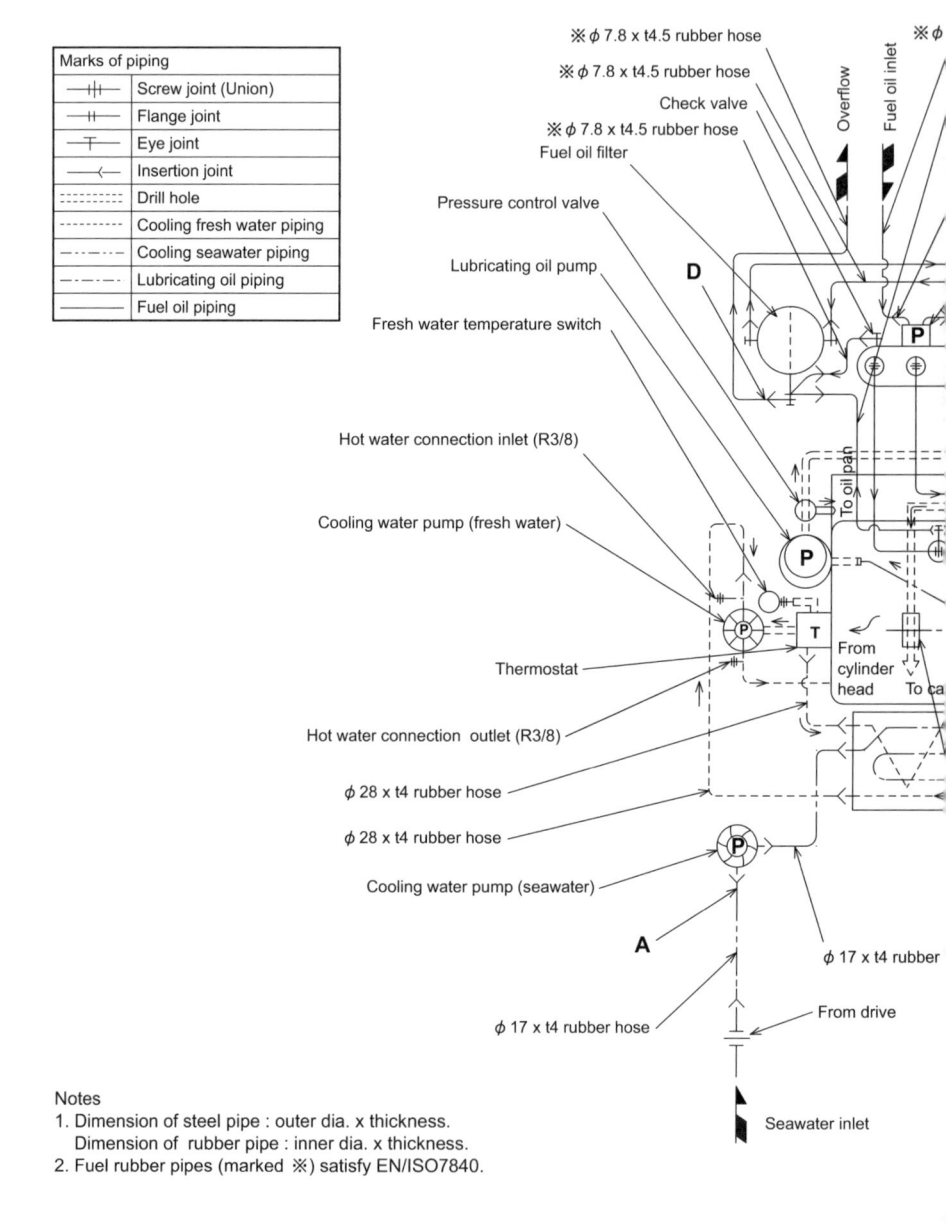

Notes
1. Dimension of steel pipe : outer dia. x thickness.
   Dimension of rubber pipe : inner dia. x thickness.
2. Fuel rubber pipes (marked ※) satisfy EN/ISO7840.

1. General

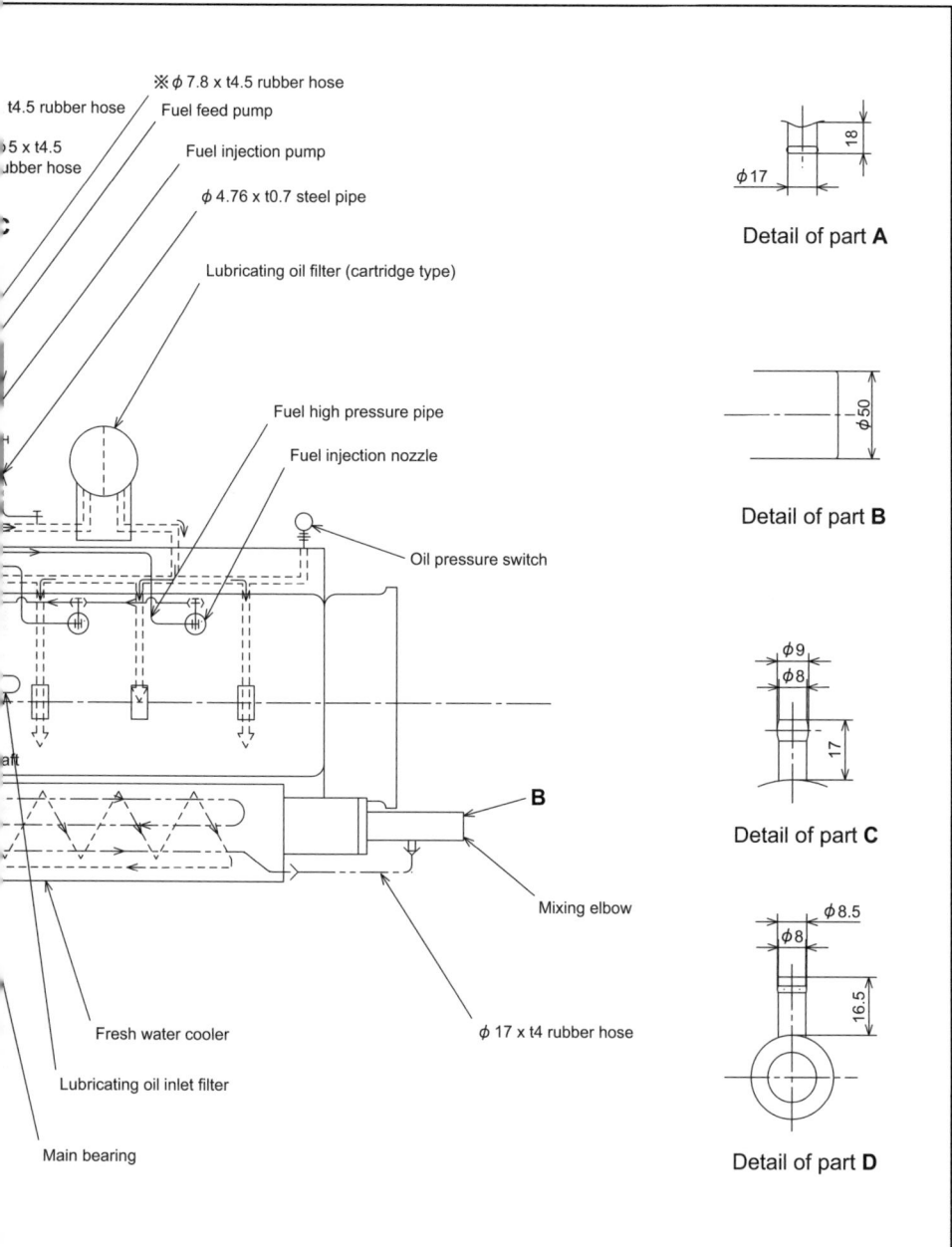

(3) 2YM15 (with KM2P-1 marine gear)

Marks of piping	
—‖‖—	Screw joint (Union)
—‖—	Flange joint
—⊤—	Eye joint
—←—	Insertion joint
======	Drill hole
— — — —	Cooling fresh water piping
— - - —	Cooling seawater piping
— - — -	Lubricating oil piping
————	Fuel oil piping

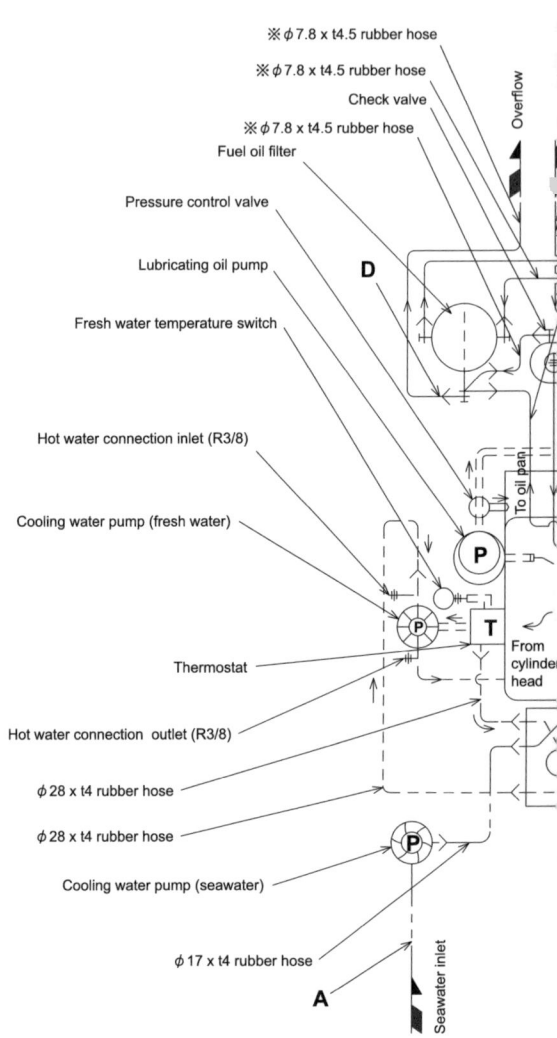

※ φ7.8 x t4.5 rubber hose
※ φ7.8 x t4.5 rubber hose
Check valve
※ φ7.8 x t4.5 rubber hose
Fuel oil filter
Pressure control valve
Lubricating oil pump
Fresh water temperature switch
Hot water connection inlet (R3/8)
Cooling water pump (fresh water)
Thermostat
Hot water connection outlet (R3/8)
φ28 x t4 rubber hose
φ28 x t4 rubber hose
Cooling water pump (seawater)
φ17 x t4 rubber hose
A
Seawater inlet
Overflow
To oil pan
From cylinder head
D
P

Notes
1. Dimension of steel pipe : outer dia. x thickness.
   Dimension of rubber pipe : inner dia. x thickness.
2. Fuel rubber pipes (marked ※) satisfy EN/ISO7840.

1. General

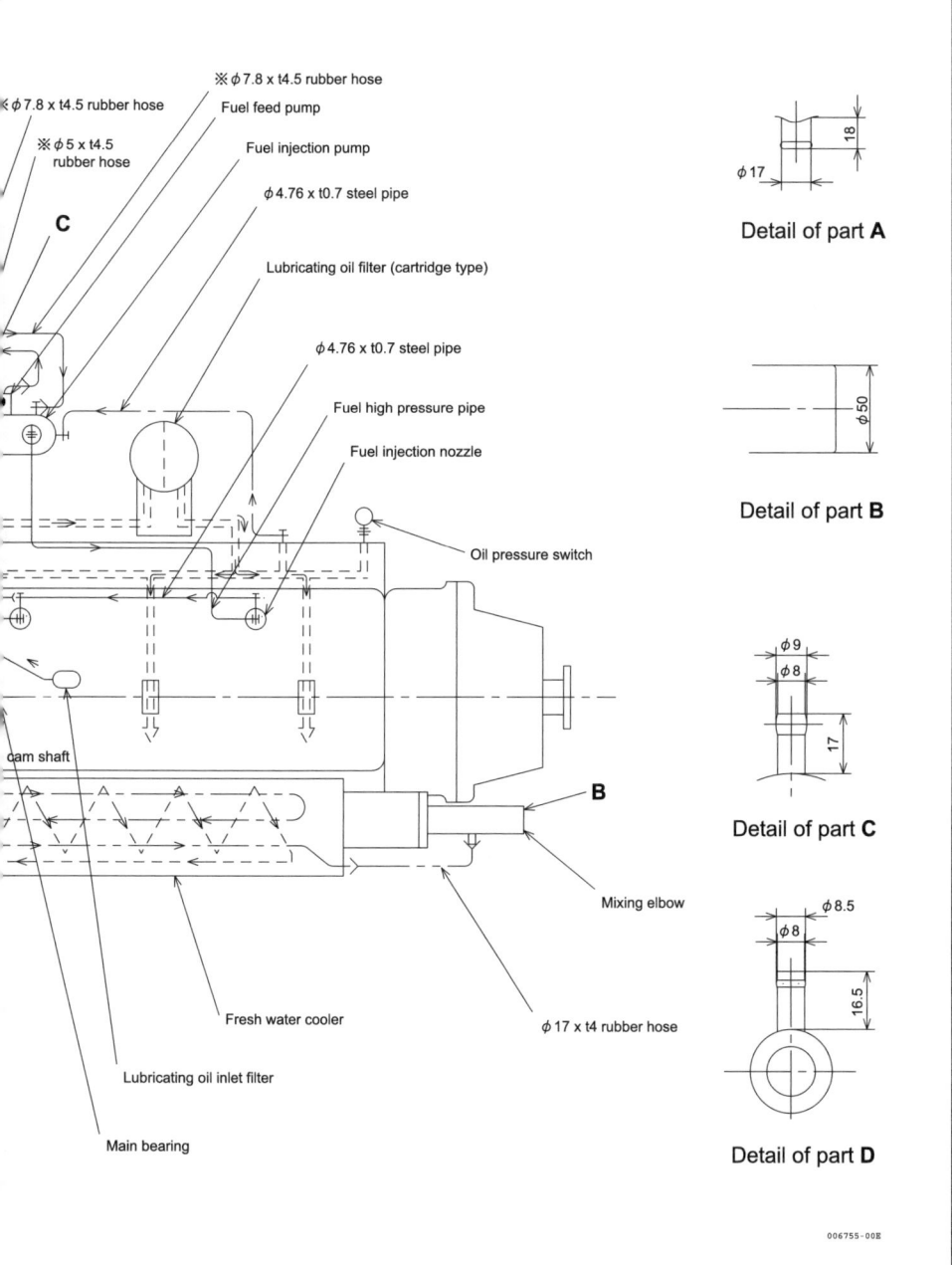

(4) 2YM15C (with sale drive SD20)

MARKS OF PIPING	
—╫—	SCREW JOINT (UNION)
—╂—	FLANGE JOINT
—┬—	EYE JOINT
—←—	INSERTION JOINT
┄┄┄┄	DRILL HOLE
┄┄┄┄	COOLING FRESH WATER PIPING
— — —	COOLING SEAWATER PIPING
—·—·—	LUBRICATING OIL PIPING
———	FUEL OIL PIPING

※ φ7.8 x t4.5 RUBBER HOSE
※ φ7.8 x t4.5 RUBBER HOSE
CHECK VALVE
※ φ7.8 x t4.5 RUBBER HOSE
FUEL OIL FILTER
PRESSURE CONTROL VALVE
LUBRICATING OIL PUMP
FRESH WATER TEMPERATURE SWITCH
HOT WATER CONNECTION INLET (R3/8)
COOLING WATER PUMP (FRESH WATER)
THERMOSTAT
HOT WATER CONNECTION OUTLET (R3/8)
φ28 x t4 RUBBER HOSE
φ28 x t4 RUBBER HOSE
COOLING WATER PUMP (SEA WATER)

OVERFLOW
D
TO OIL PAN
P
T
FROM CYLINDER HEAD
P
A
φ17xt4
FROM D
φ17 x t4 RUBBER HOSE
SEA WATER INL

NOTES
1. DIMENSION OF STEELPIPE:OUTER DIA. x THICKNESS
   DIMENSION OF RUBBER PIPE:INNER DIA. x THICKNESS
2. FUEL RUBBER PIPES(MARKED※) SATISFY EN/ISO7840.

1. General

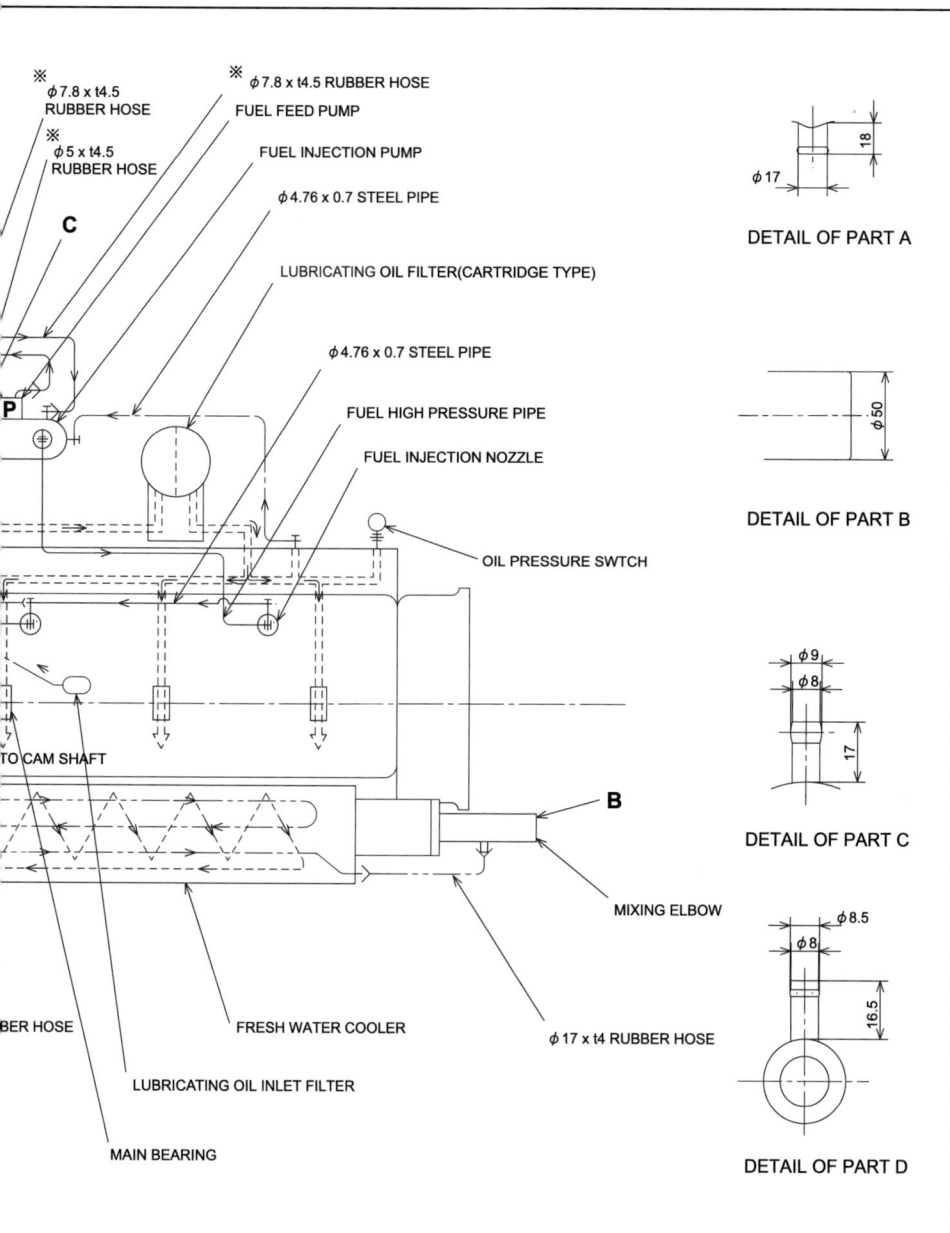

1. General

1.6 Exhaust gas emission regulation

3YM30/3YM20/2YM15 series engines are applicable with Off-road Compression Ignition engines regulations of the EPA and ARB (,California Air Resources Board) in USA and BSO regulations in Europe.

1.6.1 Engine identification (EPA/ARB)

With the regulations on exhaust gas emission worldwide, it has become necessary to identify engines in a manner to determine which regulations they comply with, hence

(1) **Emission control label (EPA and ARB)**

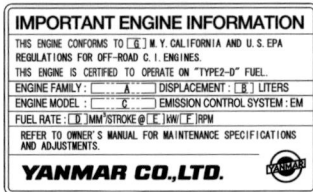

(EPA and ARB label)

(Note) Emission Control is accomplished through Engine Modification (EM-Design)

- The tamper resistance device is installed with EPA/ARB certified 3YM30/3YM20/2YM15 series engines to prevent illegal change of fuel injection volume and high idling speed.
  (Fuel injection volume : cap type, High idling speed : cap type)
- Engine family name as assigned by EPA/ARB identifying engine family group
  5YDXM1.11P3N and this identifies

YYDXM1.11P3N and this identifies

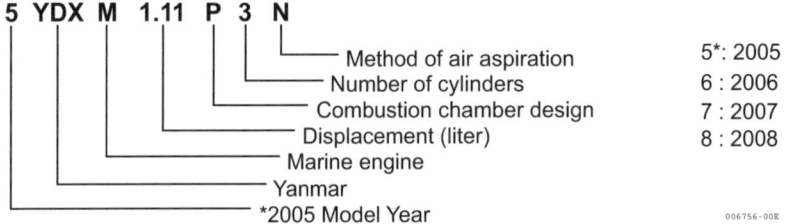

(2) Label location:

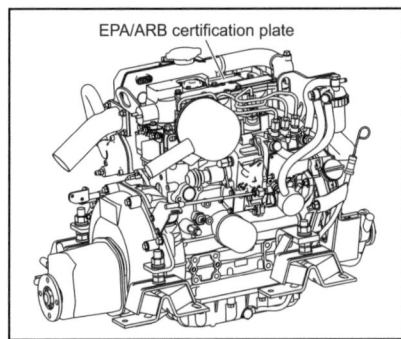

12R1

1.6.2 Exhaust gas emission standard (EPA/ARB)

Engine Power	Tier	Model Year	NOx	HC	NMHC + NOx	CO	PM
8 <= kW < 19 (11 <= hp < 25)	Tier 1	2000	-	-	9.5 (7.1)	6.6 (4.9)	0.80 (0.60)
	Tier 2	2005	-	-	7.5 (5.6)	6.6 (4.9)	0.80 (0.60)
19<= kW < 37 (25 <= hp < 50)	Tier 1	1999	-	-	9.5 (7.1)	5.5 (4.1)	0.80 (0.60)
	Tier 2	2004	-	-	7.5 (5.6)	5.5 (4.1)	0.60 (0.45)

Note
1. The transit smoke (ACC/LUG/PEAK) is not applicable.
2. The EPA recommended fuel is used.
3. The ARB emission standard is the same as the EPA's.
4. As for Model year, the year, which a regulation is applicable to, is shown.

1.6.3 Guarantee conditions for emission standard (EPA/ARB)

In addition to making sure that these conditions are met, check for any deterioration that may occur before the required periodic maintenance times.

(1) Requirement on engine installation condition

(a) Air intake negative pressure

kPa (mmAq)

Permissible
≦3.9 (400)

(b) Exhaust gas back pressure

kPa (mmAq)

Permissible
≦14.7 (1500)

(2) Fuel oil and lubricating oil

(a) Fuel: The diesel fuel oil [BS 2869 A1 or A2 (Cetane No.45 min.)]
(b) Lube oil : API grade, class CD

(3) Do not remove the caps restricting injection quantity and engine speed.

(4) Perform maintenance without fail.
Note: Inspections to be carried out by the user and by the maker are divided and set down in the "List of Periodic Inspection" on the operation manual and should be checked carefully.

EPA allows to apply the maintenance schedule for the emission related parts as follows.

Power Rating \ Parts	Maintenance period	
	Fuel nozzle cleaning	Adjustment, cleaning, repairs for fuel nozzle, fuel pump, and electronic control unit etc.
kW < 19	Every 1500 hours	Every 3000 hours
19 ≦ kW < 37	Every 1500 hours	Every 3000 hours

1. General

(5) Quality guarantee period for exhaust emission related parts

For exhaust emission related parts, follow the inspections outlined in the "List of Periodic Inspections", on the operation manual, and use the table below to carry out inspections based on operation hours or time in years. Whichever comes first is the guarantee period.

Parts Power Rating	For fuel nozzle, fuel pump
kW < 19	1500 hours or 2 years
19 ≦ kW < 37	3000 hours or 5 years

The specific emissions-related parts are

(a) Fuel injection nozzle

(b) Fuel injection pump

2. Inspection and adjustment

2.1 Periodic maintenance schedule

The engine periodic inspection timing is hard to determine as it varies with the application, load status, qualities of the fuel and lubricating oils used and handling status. General rules are described here.

○ : User-maintenance ◎ : Parts replacement ● : Shop-inspection

System	Item		Before starting	[2] Initial 50hrs. or one month	[2] Every 50hrs. or one month	[2] Every 100 hrs. or six months	[2] Every 150 hrs. or one year	[2] Every 250 hrs. or one year	[2] Every 1000 hrs. or 4 years
Whole	Visual inspection of engine outside		○						
Fuel system	Check the fuel level, and refill		○						
	Drain the fuel tank			○				○	
	Drain the fuel filter					○			
	Replace the fuel filter							◎	
	Check the injection timing								●
	Check the injection spray condition								●[*1]
Lubricating system	Check the lube oil level	Crankcase	○						
		Marine gear	○						
	Replace the lube oil	Crankcase		◎				◎	
		Marine gear		◎				◎	
		Sail drive		◎		◎			
	Replace the engine lube oil filter.			◎				◎	
Cooling system	Seawater outlet		○ During operation						
	Check cooling water level		○						
	Check the impeller of the cooling water pump (seawater pump)							○	◎
	Replace the fresh water coolant		Every year When long life coolant of a specified type is used, replacement period of two years can be obtained.						
	Clean & check the water passages								●

[*1] For EPA requirements see also 1.5 in chapter 1.
[*2] Whichever comes first

2. Inspection and adjustment

○ : User-maintenance ◎ : Parts replacement ● : Shop-inspection

System	Item	Before starting	*2 Initial 50hrs. or one month	*2 Every 50hrs. or one month	*2 Every 100 hrs. or six months	*2 Every 150 hrs. or one year	*2 Every 250 hrs. or one year	*2 Every 1000 hrs. or 4 years
Air intake and exhaust system	Clean the element of the air intake silencer						○	
	Clean the exhaust/water mixing elbow						○	
	Check the exhaust gas condition	○ During operation						
	Diaphragm assembly inspection							●
Electrical system	Check the alarm lamps & devices	○						
	Check the electrolyte level in the battery				○			
	Adjust the tension of the alternator driving belt		○				○	◎
	Check the wiring connectors						○	
Cylinder head, etc.	Check for leakage of water and oil	○ After starting						
	Retighten all major nuts and bolts							●
	Adjust intake/exhaust valve clearance		○					●
Remote control system, etc.	Check/adjust the remote control operation	○	○					●
	Adjust the propeller shaft alignment		○					●

*2 Whichever comes first

2.2 Periodic inspection and maintenance procedure

2.2.1 Check before starting
Be sure to check the following points before starting an engine every day.

No.	Inspection Item
(1)	Visual inspection of engine outside
(2)	Check the fuel level, and refill
(3)	Check the lube oil level (Crankcase/Marine gear)
(4)	Seawater outlet
(5)	Check cooling water level
(6)	Check the alarm lamps & devices
(7)	Check the leakage of water, lube oil and fuel
(8)	Check/adjust the remote control operation

(1) Visual inspection of engine outside
If any problem is found, do not use before the engine repairs have been completed.
 • Oil leak from the lubrication system
 • Fuel leak from the fuel system
 • Cooling water leak from the cooling water system
 • Damaged parts
 • Loosened or lost bolts
 • Fuel, coolant tank rubber hoses, V belt cracked, loosened clamp

(2) Check the fuel level, and refill
Check the remaining fuel oil level in the fuel tank and refill the recommended fuel if necessary.

(3) Check the lube oil level (Crankcase/Marine gear)
 1) Checking engine lube oil level
    a) Check the lube oil level of a engine with a dipstick. Insert the dipstick fully and check the oil level. The oil shall not be contaminated heavily and have appropriate viscosity. No cooling water or diesel fuel shall be mixed

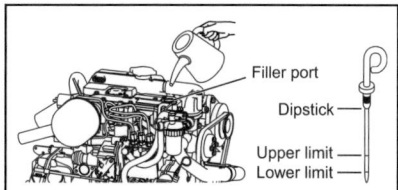

Filler port
Dipstick
Upper limit
Lower limit

Standard
The level shall be between the upper and lower limit lines on the dipstick.

Unit : liter (quart)

Model	Rake angle	Engine oil capacity (Full)
3YM30 with KM2P-1	8 degree	$2.8\,^{0}/_{-0.2}$ (3.0)
3YM20 with KM2P-1		$2.7\,^{0}/_{-0.2}$ (2.9)
2YM15 with KM2P-1		$2.0\,^{0}/_{-0.2}$ (2.1)
3YM30C with SD20	0 degree	$2.5\,^{0}/_{-0.2}$ (2.6)
3YM20C with SD20		$2.4\,^{0}/_{-0.2}$ (2.5)
2YM15C with SD20		$1.8\,^{0}/_{-0.2}$ (1.9)

 b) If the remaining engine oil level is low, fill the oil pan with the specified engine oil to the specified level through the filler port.

[NOTICE]
The engine oil should not be overfilled to exceed the upper limit line. If engine oil is overfilled, the engine may intake the engine oil in the combustion chamber during the operation, and white smoke, oil hummer or urgent rotation may occur, because the blowby gas is reduced in the suction air flow.

2. Inspection and adjustment

2) Checking marine gear lube oil level

a) Check the lube oil level of the marine gear with a dipstick.

Unit: liter (pint)

Marine gear oil capacity	
KM2P-1	Full : 0.30 (0.64)

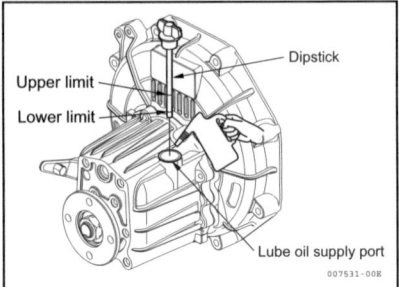

b) When the level is low, remove a oil filler cap at the top of the housing, and fill it with marine gear-clutch- lube oil to the upper limit on the dipstick.

c) Tighten the oil filler cap securely by hand.

(4) Seawater outlet
Check whether seawater comes out just after the engine has started.
If seawater doesn't come out, shut down the engine immediately.
Check the leakage of seawater in the seawater pass and the damage of the seawater pump impeller.

(5) Check cooling water level
Daily inspection of cooling water should be done only by coolant recovery tank.

- Never open the filler cap while the engine is still hot. Steam and hot water will spurt out and seriously burn you. Wait until the engine is cooled down after the engine stopped, wrap the filler cap with a rag piece and turn the cap slowly to gently release the pressure inside the flesh water tank.
- Securely tighten the filler cap after checking the flesh water tank. If the cap is tightened loosely, steam can spurt out during operation.

1) Checking cooling water volume
Check the cooling water level in the coolant recovery tank. If the water level is close to the LOW mark, open the coolant recovery tank cap and replenish the coolant recovery tank with clean soft water to the FULL mark.

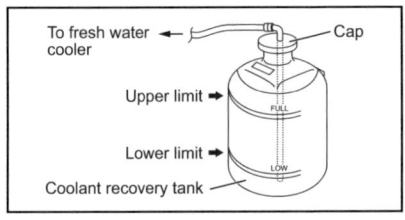

Standard
The water level of the coolant recovery tank shall be between the upper and lower limit lines.

2) Replenishing engine with water
If the cooling water leveling the coolant recovery tank is lower than the LOW mark, open the filler cap and check the cooling water level in the coolant tank. Replenish the engine with the cooling water, if the level is low.
- Check the cooling water level when the engine is cool.
  Checking when the engine is hot is dangerous. And the water volume is expanded due to the temperature.
- Daily cooling water level check and replenishing shall be done only at the coolant recovery tank.
  Usually do not open the filler cap to check or replenish.

Standard

Cooling water volume   Unit : liter (quart)

Model	Engine	Coolant recovery tank
3YM30	4.9 (5.2)	0.8 (0.8)
3YM20	4.1 (4.3)	0.8 (0.8)
2YM15	3.0 (3.2)	0.8 (0.8)

2. Inspection and adjustment

IMPORTANT:
If the cooling water runs short quickly or when the coolant tank runs short of water with the coolant recovery tank level unchanged, water may be leaking or the air tightness may be lost. Increase in the water level of the coolant recovery tank during operation is not abnormal.
The increased water in the coolant recovery tank returns to the coolant tank when the engine is cooled down.
If the water level is normal in the coolant recovery tank but low in the coolant tank, check loosened clamping of the rubber hose between the coolant tank and coolant recovery tank or tear in the hose.

(6) Check the alarm lamps & devices

Before and after starting the engine, check to see that the alarm functions normally. Failure of alarm cannot warn the lack of the engine oil or the cooling water. Make it a rule to check the alarm operation before and after starting engine every day.
When the sensor detects a problem during operation, the lamp comes on and the buzzer goes off.
Alarm lamps are located on the panel, buzzer is located on the back of panel.
Under normal conditions, the monitors are off. When there is a problem, the monitors light up.

	Battery low charge alarm	When the alternator output is too low, the lamp will come on. When charge begins, the lamp will turn off. (The alarm buzzer will not sound, when the lamp comes on.)
	Coolant high temperature alarm	When the temperature reaches the maximum (95°C [203F] or higher), the lamp will light. Continuing operation at temperatures exceeding the maximum limit will result in damage and seizure. Check the load and the fresh water cooling system for any abnormalities.
	Lubricating oil low pressure alarm	When the lube oil pressure falls below normal, the oil pressure sensor will register this and the lamp will come on and alarm will sound. Continuing operation with insufficient oil pressure will result in damage and seizure. Check the oil level.
	Water in sail drive seal alarm	When seawater is detected between the seals of sail drive, the lamp will come on and the alarm will sound.

Normal action of alarm devices

Alarm devices act as shown below. Check that the alarm lamps and buzzer are working normally, when the key is turned on.

Key switch		OFF → ON	START → ON
Engine		Before start	Running
Alarm buzzer		Goes on	Stop
Alarm lamps	Battery low charge alarm	Light	Off
	Coolant high temperature alarm	Off	Off
	Lube oil low pressure alarm	Light	Off
	Water in sail drive seal alarm	Off	Off

(7) Check the leakage of water, lube oil and fuel.

Before and after starting the engine, check the leakage of cooling water and seawater from cooling water system. Also check the leakage of lube oil and fuel.

(8) Check/adjust the remote control operation

Make sure that the accelerator of a boat can be operated smoothly before starting the engine. If it feels heavy to manipulate, lubricate the accelerator cable joints and pivots. Adjust the accelerator cable if there is a dislocation or excessive play between the accelerator and the governor lever.

2. Inspection and adjustment

2.2.2 inspection after initial 50 hours or one month operation

Be sure to check the following points after initial 50 hours or one month operation, whichever comes first.

No.	Inspection Item
(1)	Drain the fuel tank
(2)	Replace the engine lube oil and the lube oil filter
(3)	Replace the marine gear & sail drive lube oil
(4)	Adjust the tension of the alternator driving belt
(5)	Adjust the intake/exhaust valve clearance
(6)	Check/adjust the remote control operation
(7)	Adjust the propeller shaft alignment

(1) Drain the fuel tank

1) Put a pan under the drain cock to catch the fuel.
2) Open the drain cock and drain off any water or dirt collected.
3) When the water and dirt are drained off and the fuel comes out, close the drain cock.

(2) Replace the engine lube oil and the lube oil filter.

During the operation of an engine, the oil is quickly contaminated due to the initial wear of internal parts. The lube oil must therefore be replaced early. It is easiest and most effective to drain the engine lube oil after operation while the engine is still warm.
Replace the lube oil filter at the same time.

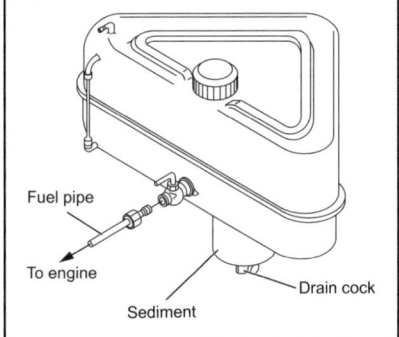

 Beware of oil splashes if extracting the Lube oil while it is hot

1) Remove the lube oil dipstick and also the oil filler cap at the top of the rocker arm cover. Attach the oil drain pump to the dipstick guide and drain off the lube oil.

[Notice]
For easier draining, remove the oil filler cap (yellow) at the top of the rocker arm cover.
When lube oil is absorbed without removing a oil filler cap, negative pressure grows big in the crankcase and it may cause the rubber of the diaphragm cracked.

2) Turn the lube oil filter counter-clockwise using a filter wrench to remove it.

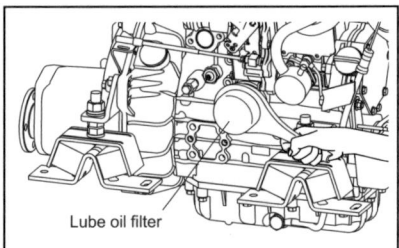

2. Inspection and adjustment

3) Moisten the new oil filter gasket with the lube oil and install a new lube oil filter manually turning it clockwise until it comes into contact with the mounting surface, and tighten it further to 3/4 of a turn with the filter wrench.
Tightening torque: 20-24N•m (177-212 lb-in)

Applicable oil filter Part No.
119305-35150

4) Fill with new lube oil.
5) Perform a trial run of the engine and check the oil leakage.
6) Approximately 10 minutes after stopping the engine, check the oil level by using the oil dipstick. Add oil if the level is too low.

Standard
The level shall be between the upper and lower limit lines on the dipstick.

Unit : liter (quart)

Model	Rake angle	Engine oil capacity (Full)
3YM30 with KM2P-1	8 degree	2.8 $^0/_{-0.2}$ (3.0)
3YM20 with KM2P-1		2.7 $^0/_{-0.2}$ (2.9)
2YM15 with KM2P-1		2.0 $^0/_{-0.2}$ (2.1)
3YM30C with SD20	0 degree	2.5 $^0/_{-0.2}$ (2.6)
3YM20C with SD20		2.4 $^0/_{-0.2}$ (2.5)
2YM15C with SD20		1.8 $^0/_{-0.2}$ (1.9)

[Notice]
When checking the lube oil level right after the engine running, the oil level in the dipstick guide decreases drastically, and the accurate oil measurement can't be performed because the pressure in the cylinder block decreases with the function of the diaphragm in the rocker arm cover. Therefore, measure the lube oil level about 10 minutes later after stopping the engine or after removing the oil filler cap.

Note:
The yellow label shown right is attached on the dipstick.

OIL CHECK 10 MINUTES AFTER ENGINE STOP.

(3) Replace the marine gear lube oil.
During initial operation, the oil is quickly contaminated due to the initial wear of internal parts. The lube oil must therefore be replaced early.
1) Remove the cap from the filler port and attach the oil drain pump. Drain off the oil.
2) Fill with new lube oil.

Standard

Unit : liter (pint)

Marine gear oil capacity	
KM2P-1	Full: 0.30 (0.64)

3) Perform a trial run of the engine and check the oil leakage.
Refer to the sail drive operation manual for the procedure of replacing the sail drive lube oil.

Unit : liter (pint)

Sail drive oil capacity	
SD20	Full: 2.2 (4.7)

2. Inspection and adjustment

(4) Adjust the tension of the alternator drive belt

When there is not enough tension in the V-belt, it will slip and the cooling water pump will fail to supply cooling water. Engine over-heating and the seizure will occur.

When there is too much tension in the V-belt, the belt will become damaged more quickly and the bearing of the cooling water pump may be damaged.

Check and adjust the V-belt tension (deflection) in the following manner.

[Notice]
Be especially careful not to splash engine oil on the V-belt, because it will cause slipping, stretching and aging of the belt.

1) Remove the belt cover. Check the tension of the V-belt by pressing down on the middle of the belt with your finger [approx. 98N(10kgf)].

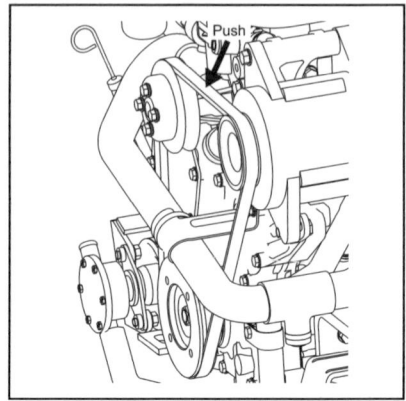

The specified deflection should be as follows.

For used V-belt	8-10 mm (0.315-0.393 inches)
For new V-belt	6-8 mm (0.236-0.315 inches)

- "New V-belt" refers to a V-belt which has been used less than 5 minutes on a running engine.
- "Used V-belt" refers to a V-belt which has been used on a running engine for 5 minutes or more.

2) If necessary, adjust the V-belt tension (deflection). To adjust the V-belt tension, loosen the set bolt for the belt adjuster and move the alternator to tighten the V-belt.

3) Visually check the V-belt for cracks, oiliness or wear. If any, replace the V-belt with new one.

[Notice]
When the V-belt will be replaced with new one, loosen the set bolt, move the alternator and also loosen the V-pulley set bolts for the cooling water pump. Remove the V-belt.

After replacing with a new V-belt and adjusting the tension, run the engine for 5 minutes and readjust the deflection to the value in the table above.

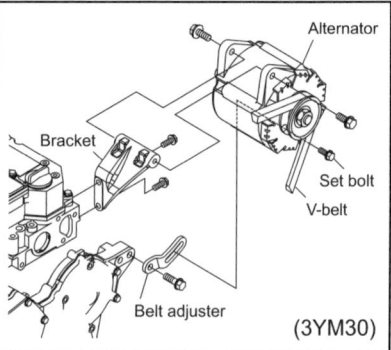

(3YM30)

2. Inspection and adjustment

(5) Adjusting the intake/exhaust valve clearance
Make measurement and adjustment while the engine is cold.

1) Valve clearance measurement
   a) Remove the rocker arm cover above cylinder head.
   b) Set the No.1 cylinder in the compression TDC
      Turn the crankshaft to bring the piston of the No.1 cylinder to its compression top dead center while watching the rocker arm motion, the timing mark of the flywheel housing and the top mark of the flywheel.
      (Position where both the intake and exhaust valves are closed.)

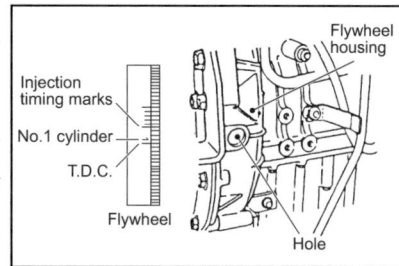

(Notes)
- The crankshaft shall be turned clockwise as seen from the gear case side.
- The No.1 cylinder position is on the opposite side of the gear case.
- Since there is a clearance between the rocker arm and valve at the compression top dead center, the position of TDC can be checked by hand. Also see that the top mark on the flywheel aligns with the mark on the flywheel housing. If there is no valve clearance, disassemble and inspect around the valve seat, since the valve seat may be worn abnormally.

   c) Valve clearance measurement
      Insert a thickness gage between the rocker arm and valve cap, and record the measured valve clearance.
      (Use it as the data for estimating the wear state.)
   d) Adjusting other cylinders
      Turn the crankshaft 240° and make adjustment for the No.3 cylinder. Then adjust the No.2 cylinder in this order.
      The cylinder to be adjusted first does not have to be the No.1 cylinder. Select and adjust the cylinder where the piston is the nearest to the top dead center after turning, and make adjustment for other cylinders in the order of ignition by turning the crankshaft 240° each time.

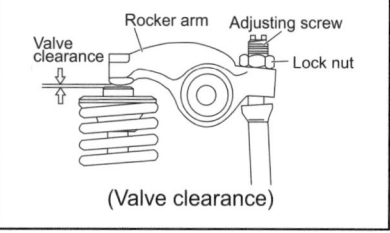

(Valve clearance)

The adjustment method of reducing the flywheel turning numbers (for reference):
Set No.1 cylinder to the compression T.D.C. and adjust the clearance of the ● mark of the below table. Next, turn the flywheel once (the suction/ exhaust valve of No.1 cylinder is in the position of the overlap T.D.C. at this time), and adjust the clearance of the ○ mark.
Ignition order of 3 cylinder engines: 1 → 3 → 2

Cylinder No.	1		2		3		
Valve	Suction	Exhaust	Suction	Exhaust	Suction	Exhaust	
No.1 compression T.D.C	●	●	●			●	The first time
No.1 overlap T.D.C				○	○		The second time

2) Valve clearance inspection and adjustment
   a) Loosen adjusting bolts
      Loosen the lock nut and adjust the screw. And check the valve for any inclination of valve cap, entrance of dirt or wear.

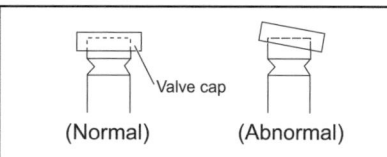

(Normal)   (Abnormal)

2. Inspection and adjustment

b) Measure valve clearance
   Insert a 0.2 mm thickness gage between the rocker arm and valve cap, and adjust the valve clearance. Tighten the adjusting bolt.

Standard valve clearance (mm)
0.15-0.25

c) Apply oil to the contact surface between adjusting screw and push rod.

d) Adjusting other cylinders
   Turn the crankshaft 240° then and make adjustment for the No.3 cylinder. Then adjust the No.2 cylinder in this order.
   The cylinder to be adjusted first does not have to be the No.1 cylinder. Select and adjust the cylinder where the piston is the nearest to the top dead center after turning, and make adjustment for other cylinders in the order of ignition by turning the crankshaft 240° each time.

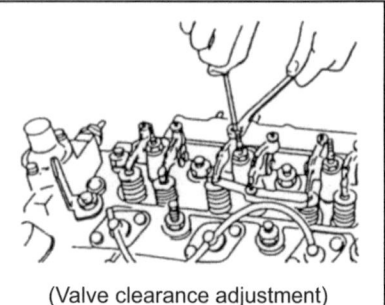
(Valve clearance adjustment)

(6) Check and adjust the remote control operation
The various control levers on the engine side are connected to the remote control lever by the remote control cable. The cable will become stretched and the attachments loose after long hours of use causing deviation. It is dangerous to control operation under these conditions, and the remote control cable must be checked and adjusted periodically.

1) Adjusting the throttle remote control cable
   Check to see that the control lever on the engine side moves to the high speed stop position and low speed stop position when the remote control lever is moved to H (high speed) and L (low speed) respectively.
   When there is deviation, loosen the clamp for the remote control cable on the engine side and adjust.
   Adjust the high speed stop position first and then adjust the low speed idling by the adjustment bolt on the remote control lever.

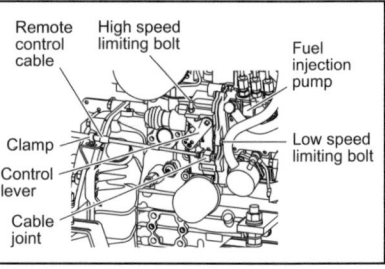

⚠ CAUTION — Never adjust the high speed limiting bolt. This will void warranty.

2) Adjusting the clutch remote control cable
   Check that the shift lever moves to the correct position, when the remote control handle is put in NEUTRAL, FORWARD and REVERSE position.
   Use the NEUTRAL position as the standard for adjustment. When there is deviation, loosen the clamp for the remote control cable and adjust the shift lever position

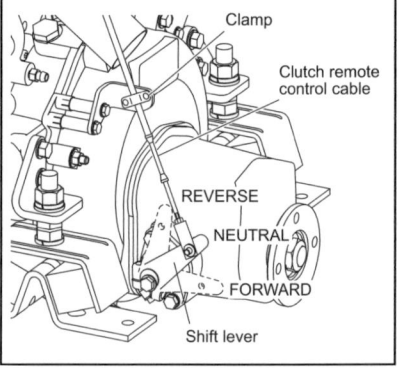

2. Inspection and adjustment

(7) Adjust the propeller shaft alignment

The flexible engine mounts is compressed a little in the initial engine operation and it may cause the centering misalignment between the engine & the propeller shaft.

1) Check unusual noise and vibration of the engine / boat hull, while increasing the engine speed gradually and lowering it.
2) If there is unusual noise and/or vibration, adjust the propeller shaft alignment. (Refer to 6.4.5 "Centering the Engine" in the installation manual for pleasure boat use.)

2.2.3 Inspection every 50 hours or monthly

Be sure to check the following points every 50 hours or monthly, whichever comes first.

No.	Inspection Item
(1)	Drain the fuel filter
(2)	Check the electrolyte level in the battery

(1) Drain the fuel filter.

1) Close the fuel cock of a fuel tank.
2) Loosen the retaining ring and remove the filter cap. Drain off any water and dirt collected inside.
3) After reassembly, be sure to vent air from the fuel system.
   If air is in the fuel system, fuel cannot reach the fuel injection pump. Vent the air in the system according to the following procedures.

Fuel system air bleeding procedures

1) Check the fuel level in the fuel tank. Replenish if Insufficient
2) Loosen the air bleeding bolt at the top of the fuel filter by turning it 2 or 3 times.
3) Feed fuel with the fuel feed pump by moving the lever on the left side of the feed pump up and down.
4) Allow the fuel containing air bubbles to flow out from the air bleeding bolt hole. When the fuel no longer contains bubbles, tighten the air bleeding bolt. This completes the air bleeding of the fuel system.
5) After the engine start-up, the automatic air-bleeding device works to purge the air in the fuel system. No manual air-venting is required for normal engine operation.

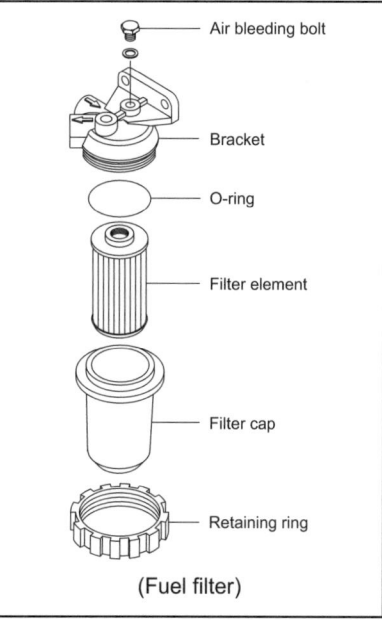

(Fuel filter)

2. Inspection and adjustment

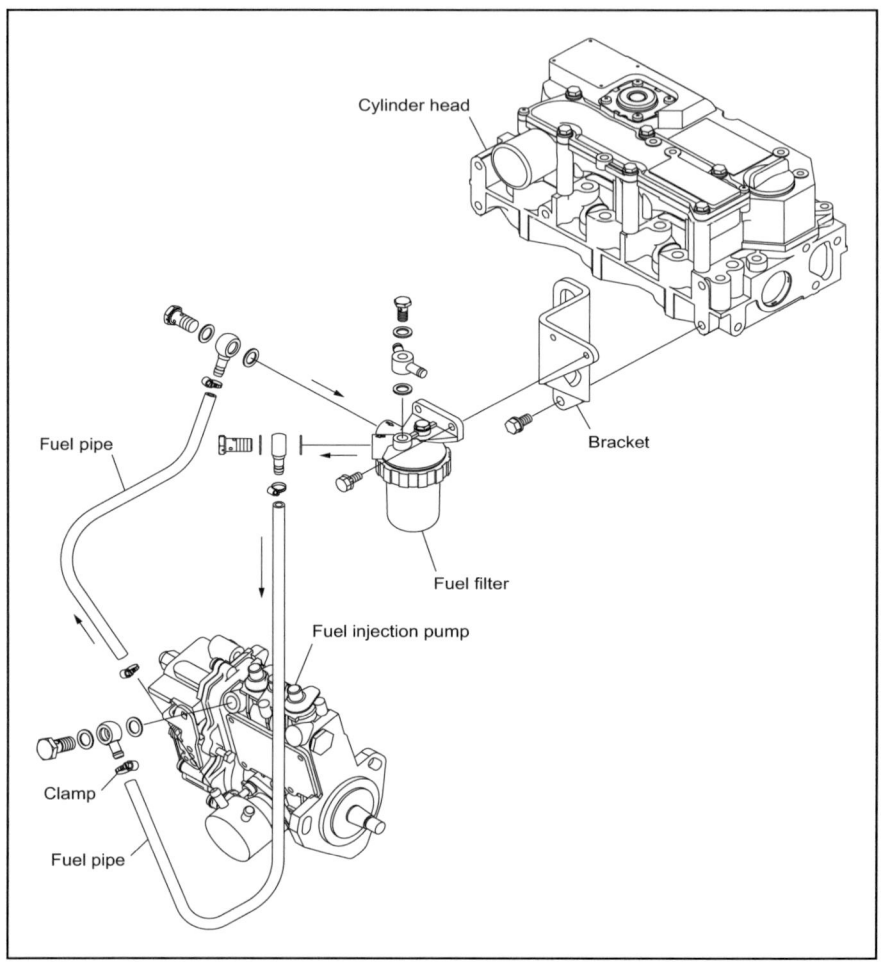

2. Inspection and adjustment

(2) Check the electrolyte level in the battery

Fire due to electric short-circuit
- Make sure to turn off the battery switch or disconnect the negative cable (-) before inspecting the electrical system. Failure to do so could cause short-circuiting and fires.
- Always disconnect the (-) Negative battery cable first before disconnecting the battery cables from battery. An accidental "Short circuit" may cause damage, fire and or personal injury.
  And remember to connect the (-) Negative battery cable (back onto the battery) LAST.

Proper ventilation of the battery area
Keep the area around the battery well ventilated, paying attention to keep away any fire source. During operation or charging, hydrogen gas is generated from the battery and can be easily ignited.

Do not come in contact with battery electrolyte
Pay sufficient attention to avoid your eyes or skin from being in contact with the fluid. The battery electrolyte is dilute sulfuric acid and causes burns. Wash it off immediately with a large amount of fresh water if you get any on you.

Battery structure

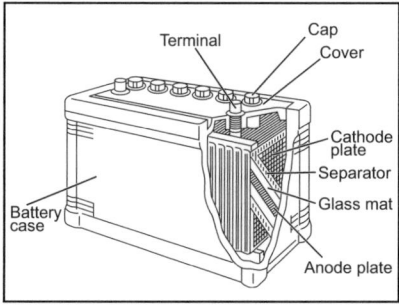

2. Inspection and adjustment

(1) Electrolyte level
- Check the level of fluid in the battery.
  When the amount of fluid nears the lower limit, fill with battery fluid (available in the market) to the upper limit. If operation continues with insufficient battery fluid, the battery life is shortened, and the battery may overheat and explode.
- Battery fluid tends to evaporate more quickly in the summer, and the fluid level should be checked earlier than the specified times.
- If the engine cranking speed is so slow that the engine does not start up, recharge the battery.
- If the engine still will not start after charging, replace the battery.
- Remove the battery from the battery mounting of the machine unit after daily use if letting the machine unit leave in the place that the ambient temperature could drop at -15°C or less. And store the battery in a warm place until the next use the unit to start the engine easily at low ambient temperature.

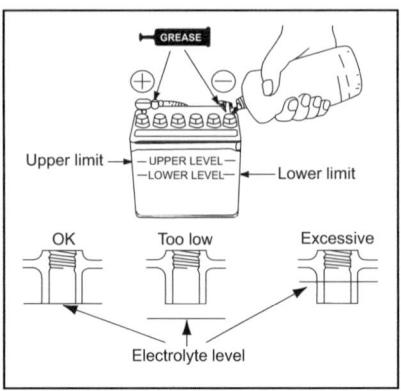

(2) Battery charge
Use a battery tester or hydrometer and check the battery condition. If the battery is discharged, recharge it.

(a) Measurement with a battery tester
When checking the battery with the batter tester, connect the red clip of the tester to the battery positive (+) terminal and black clip to the battery negative (-) terminal by pinching them securely, and judge the battery charge level from the indicator position.

Green zone: Normal
Yellow zone: Slightly discharged
Red zone: Defective or much discharged

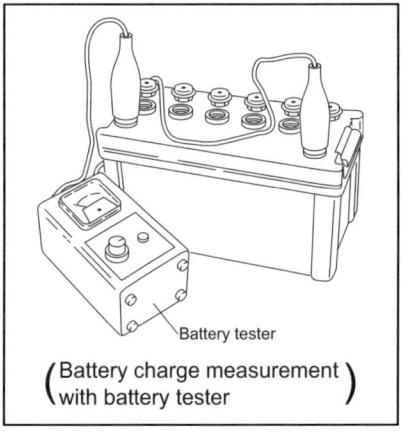

(Battery charge measurement with battery tester)

2. Inspection and adjustment

(b) Measurement with hydrometer
When using a hydrometer, the measured specific gravity must be corrected according to the temperature at the time of measurement. The specific gravity of battery electrolyte is defined with 20°C as the standard. Since the specific gravity increases or decreases by 0.0007 when the temperature varies by 1°C, correct the value according to the equation below.

$$S_{20} = St + 0.007\,(t - 20)$$

where:
- S_{20}: Converted specific gravity at 20°C
- St: Specific gravity at measurement
- t: Electrolyte temperature at measurement

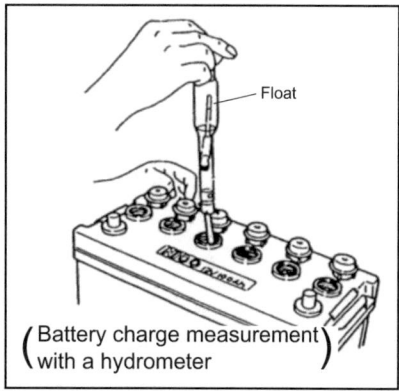
(Battery charge measurement with a hydrometer)

Specific gravity and remaining battery charge

Specific gravity (20°C)	Discharged quantity of electricity (%)	Remaining charge (%)
1.28	0	100
1.26	10	90
1.24	20	80
1.23	25	75

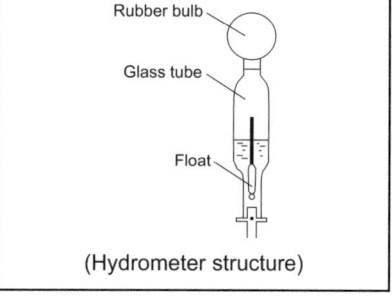
(Hydrometer structure)

(3) Terminals
Clean if corroded or soiled.

(4) Mounting bracket
Repair or replace it if corroded.
Retighten if loosened.

(5) Battery appearance
Replace the battery if cracked or deformed.
Clean with fresh water if contaminated.

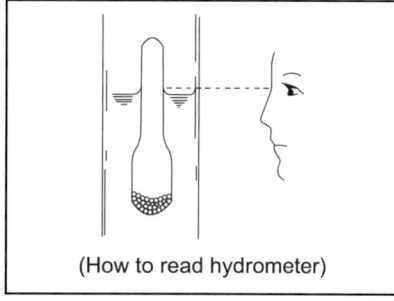
(How to read hydrometer)

2. Inspection and adjustment

2.2.4 Inspection every 100 hours six months
Be sure to check the following points every 100 hours or six months operation, whichever comes first.

No.	Inspection Item
(1)	Replace the lube oil in a sail drive.

(1) Replace the lube oil in a sail drive.
Refer to the sail drive operation manual for the procedure of replacing the oil.

2.2.5 Inspection every 150 hours or one year
Be sure to check the following points every 150 hours or one year operation, whichever comes first.

No.	Inspection Item
(1)	Replace the engine lube oil.
(2)	Replace the marine gear lube oil.

(1) Replace the engine lube oil
During the operation of an engine, the oil deteriorates by the high temperature of the engine and is contaminated due to the wear of internal parts. The lube oil must therefore be replaced periodically. It is easiest and most effective to drain the engine lube oil after operation while the engine is still warm.

⚠ CAUTION Beware of oil splashes if extracting the Lube oil while it is hot

1) Remove the lube oil dipstick and also the oil filler cap at the top of the rocker arm cover. Attach the oil drain pump to the dipstick guide and drain off the lube oil.

[Notice]
For easier draining, remove the oil filler cap (yellow) at the top of the rocker arm cover.
When lube oil is absorbed without removing a oil filler cap, negative pressure grows big in the crankcase and it may cause the rubber of the diaphragm cracked.

2) Fill with new lube oil.
3) Perform a trial run of the engine and check the oil leakage.
4) Approximately 10 minutes after stopping the engine, check the oil level by using the oil dipstick. Add oil if the level is too low.

Standard
The level shall be between the upper and lower limit lines on the dipstick.

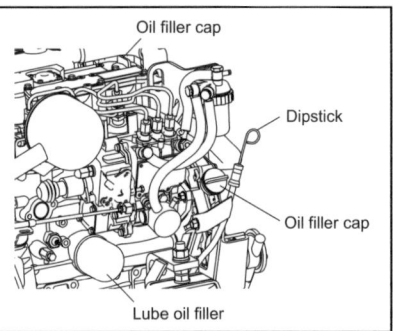
Lube oil filler

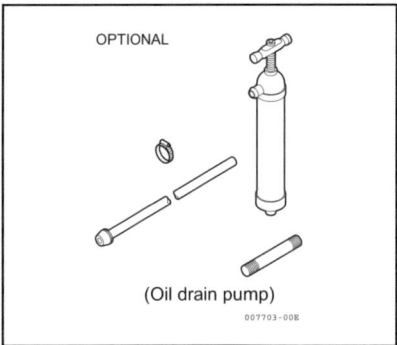
(Oil drain pump)

Unit : liter (quart)

Model	Rake angle	Engine oil capacity (Full)
3YM30 with KM2P-1	8 degree	2.8 $^0/_{-0.2}$ (3.0)
3YM20 with KM2P-1		2.7 $^0/_{-0.2}$ (2.9)
2YM15 with KM2P-1		2.0 $^0/_{-0.2}$ (2.1)
3YM30C with SD20	0 degree	2.5 $^0/_{-0.2}$ (2.6)
3YM20C with SD20		2.4 $^0/_{-0.2}$ (2.5)
2YM15C with SD20		1.8 $^0/_{-0.2}$ (1.9)

[Notice]
When checking the oil level right after the engine running, the oil level in the dipstick guide decreases drastically, and the accurate oil measurement can't be performed because the pressure in the cylinder block decreases with the function of the diaphragm in the rocker arm cover. Therefore, measure the lube oil level after removing the oil filler cap or about 10 minutes later after stopping the engine.

(2) Replace the marine gear lube oil.
Refer to 2.2.2(3) for the procedure.

2. Inspection and adjustment

2.2.6 Inspection every 250 hours or one year

Be sure to check the following points every 250 hours or one year operation, whichever comes first.

No.	Inspection Item
(1)	Drain the fuel tank
(2)	Replace the fuel filter
(3)	Replace the marine gear lube oil
(4)	Replace the engine lube oil filter
(5)	Check the impeller of the cooling water pump (seawater pump)
(6)	Replace the fresh water coolant
(7)	Clean the element of the air intake silencer
(8)	Clean the exhaust/water mixing elbow
(9)	Adjust the tension of the alternator driving belt
(10)	Check the wiring connectors

(1) Drain the fuel tank
Refer to 2.2.2(1) for the procedure.

(2) Replace the fuel filter
Replace the fuel filter with new one at the specified interval, before it is clogged with dust to adversely affect the fuel flow. Also, replace the fuel filter after the engine has fully been cooled.

1) Close the fuel cock of the fuel tank.
2) Remove the fuel filter using a filter wrench (customer procured). When removing the fuel filter, hold the bottom of the fuel filter with a piece of rag to prevent the fuel oil from dropping.
3) Clean the filter mounting surface and slightly apply fuel oil to the gasket of the new fuel filter.
4) Install the new fuel filter manually turning until it comes into contact with the mounting surface, and tighten it further to one turn using a filter wrench.

   Tightening torque: 20-24N•m

Applicable fuel filter element Part No.	O-ring Part No.
104500-55710	24341-000440

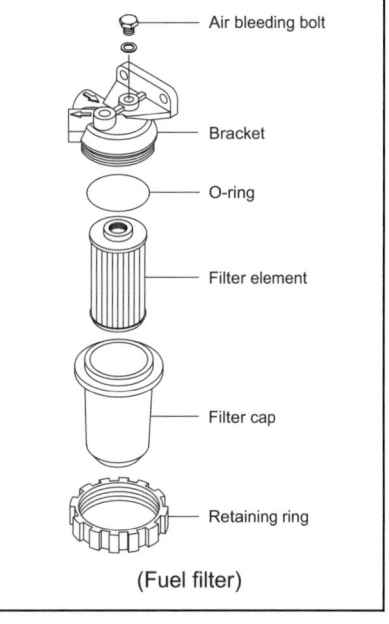

(Fuel filter)

5) Bleed the fuel system. Refer to 2.2.3.(1).

IMPORTANT:
Be sure to use genuine Yanmar part (super fine mesh filter). Otherwise, it results in engine damage, uneven engine performance and shorten engine life.

2. Inspection and adjustment

(3) Replace the marine gear lube oil.
Refer to 2.2.2(3) for the procedure.

(4) Replace the engine lube oil filter.
Refer to 2.2.2(2) for the procedure.

(5) Check the impeller of the cooling water pump (seawater pump).

1) Remove the seawater pump cover and take out the O-ring, impeller and wear plate.
2) Inspect the rubber impeller, checking for splitting around the outside, damage or cracks, and replace if necessary.
Depending on the use, the inside parts of a seawater pump may deteriorate and the discharge performance may drop. At the specified interval or when the discharge volume of seawater is reduced, inspect the seawater pump in accordance with the following procedures.

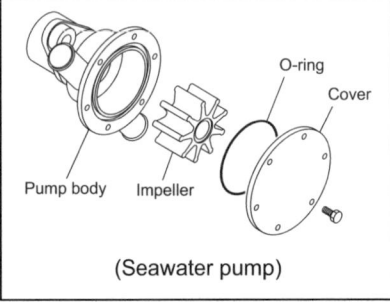
(Seawater pump)

   a) Loosen the seawater pump cover set bolts and remove the cover.
   b) Illuminate the inside of the seawater pump with a flashlight and inspect.
   c) If no damage is found, reassemble the cover.
   d) If any of the following problems are found, take out the O-ring and impeller. If necessary, replace with new one and reassemble the cover.
   • Impeller blades are cracked or nicked. Edges or surfaces of the blades are marred or scratched.

Note: The impeller must be replaced periodically (every 1000 hrs or four years whichever comes first).

   e) If a large amount of water leaks continuously from the water drain pipe beneath the seawater pump during engine operation, check the oil seal inside the seawater pump.

[Notice]
The seawater pump turns in the counterclockwise direction viewed from driving side. If the impeller has been removed for any reason and must be reassembled, be very careful not to make a mistake in impeller direction and turn it in the wrong direction. Additionally, if the engine is being turned manually, be careful to turn it in the correct direction. Incorrect turning will twist the impeller blades and damage it.

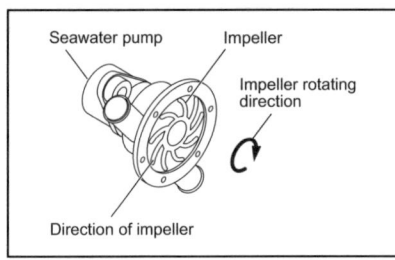

32R1

2. Inspection and adjustment

(6) Replace the fresh water coolant.

Be sure to replace the fresh water every year. When the long life coolant is used of the specified type, the replacement period of two years can be obtained.
Use clean soft water and be sure to add the Long Life Coolant Antifreeze (LLC) to the cooling water in order to prevent rust built up and freezing.
Cooling performance drops when cooling water is contaminated with rust and scale.
Even if antifreeze or antirust is added, the cooling water must be replaced periodically because the properties of the agent will degenerate.

 1) Open the two cocks for fresh water and extract the fresh water.

Note:
 One drain cock is behind the belt cover. Remove the belt cover and open the cock.

 2) Close the two cocks for fresh water.

Fresh water line	Seawater line
2	2

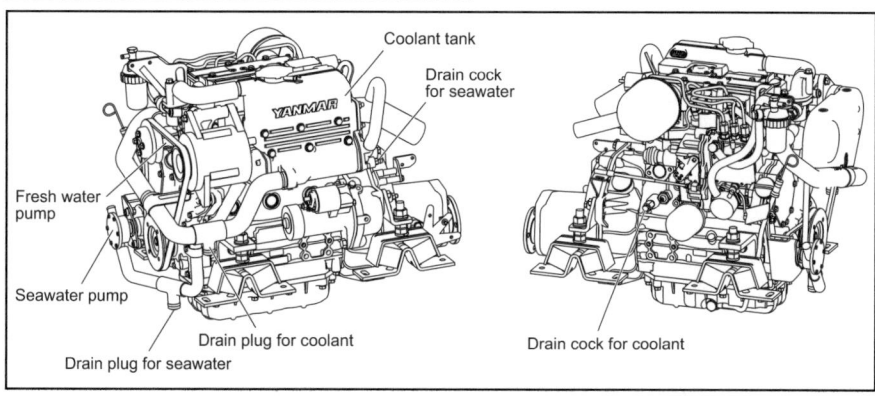

 3) Remove the filler cap of the fresh water cooler by turning the cap counterclockwise 1/3 of a turn.

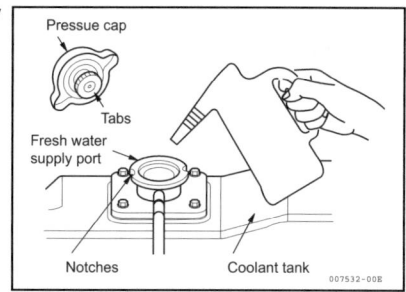

 4) Pour cooling water slowly into the coolant tank so that air bubbles do not develop. Pour until the water overflows from the filler port.

 If the filler cap is loose, hot steam and water will spout out which may cause burns.

2. Inspection and adjustment

5) After supplying cooling water, fit the filler cap and tighten it firmly. To reassemble the cap, align the tabs on the bottom of the cap with the notches on the filler port and turn clockwise 1/3 of a turn.

6) Remove the coolant recovery tank cap and fill with coolant mix to the lower limit. Replace the cap to the coolant recovery tank.

   Coolant recovery tank capacity: 0.8 L (1.7 pints)

7) Check the rubber hose connecting the coolant recovery tank to the fresh water cooler. Be sure the hose is securely connected and there is no looseness or damage.
   When the hose is not watertight, an excessive amount of cooling water will be used.

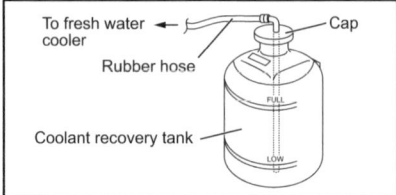

(7) Clean the element of the air intake silencer

1) Inspection of the intake silencer
   Occasionally, disassemble the intake silencer, remove the polyurethane element and inspect it. Because the element filters the air, if it is used over a long period of time it will become clogged and this decreases the amount of intake air, and may also be a cause of decreased output.

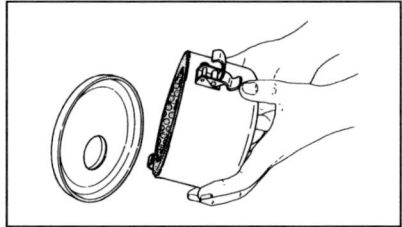

2) Washing the intake silencer element
   Wash the air intake silencer element with a neutral detergent.

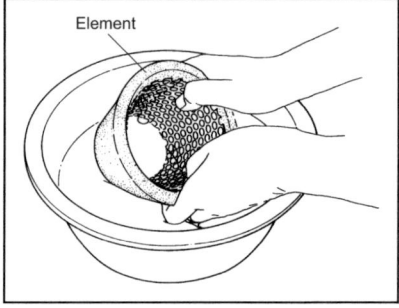

2. Inspection and adjustment

(8) Clean the exhaust/water mixing elbow.
There are two types of mixing elbows, the L-type and the U-type. The mixing elbow is attached to the exhaust manifold. The exhaust gas is mixed with seawater in the mixing elbow.

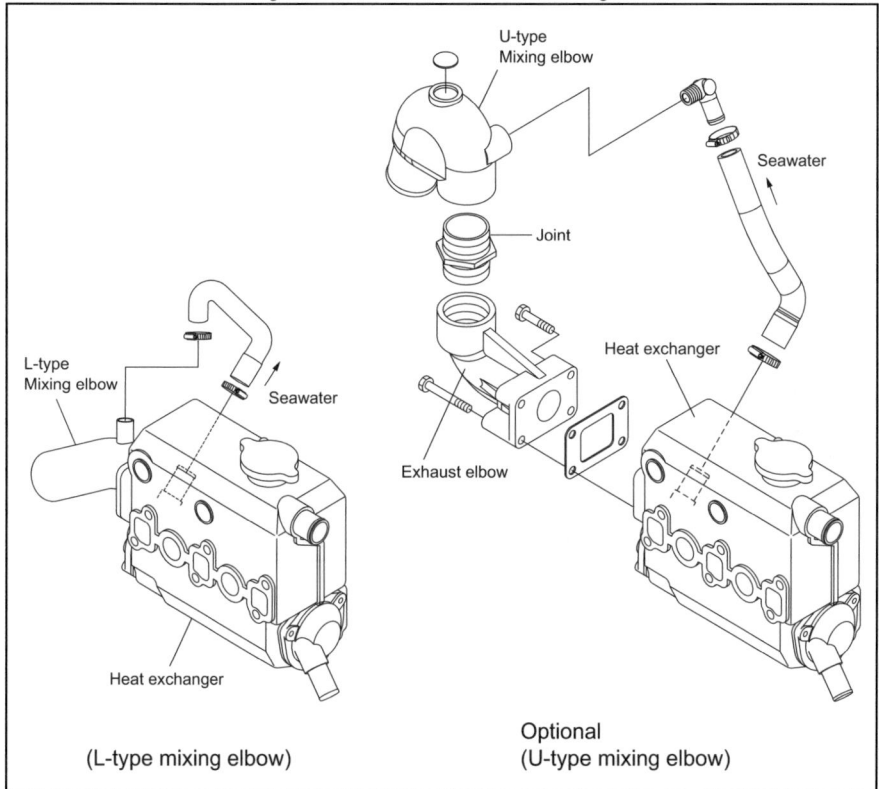

(L-type mixing elbow)

Optional
(U-type mixing elbow)

1) Clean dirt and scale out of the air pass and seawater pass of the mixing elbow.
2) Repair the crack or damage of the mixing elbow by welding, or replace if necessary.
3) Inspect the gasket and replace if necessary.

(9) Adjust the tension of the alternator driving belt.
Refer to 2.2.2(4).

(10) Check the wiring connectors.
Check whether each connection part doesn't have loosened.

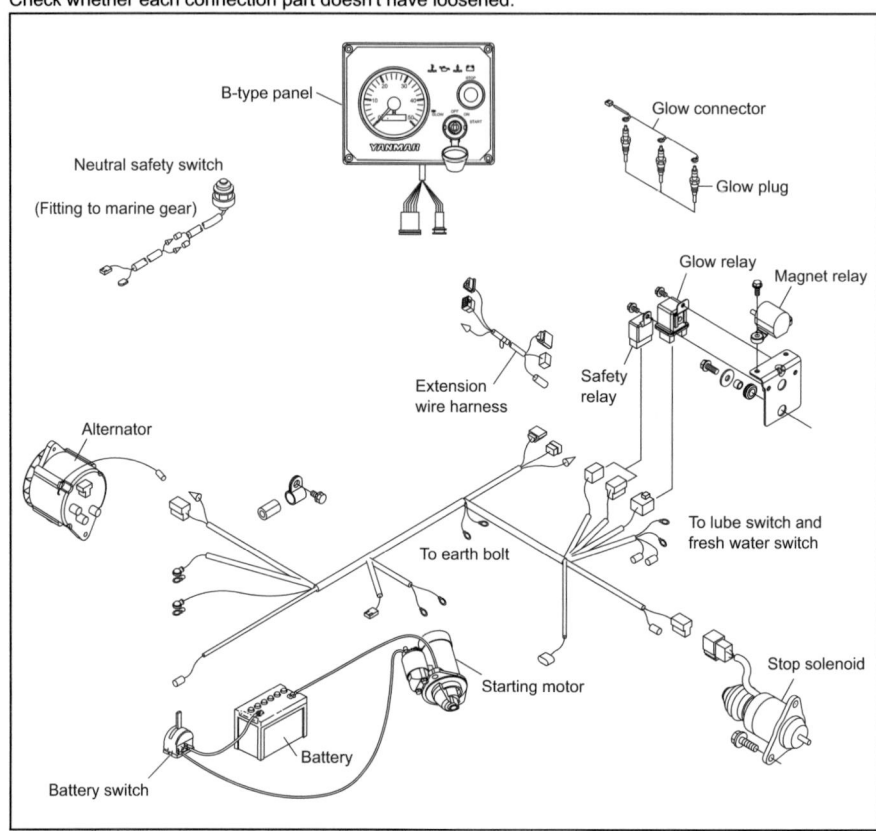

2. Inspection and adjustment

2.2.7 Inspection every 1,000 hours or four years

Be sure to check the following points every 1,000 hours or four years operation, whichever comes first.

No.	Inspection Item
(1)	Check the injection timing
(2)	Check the injection spray condition of a fuel nozzle
(3)	Check the impeller of the cooling water pump (seawater pump)
(4)	Clean & check the water passages
(5)	Diaphragm assembly inspection
(6)	Adjust the tension of the alternator driving belt
(7)	Retighten all major nuts and bolts
(8)	Adjust intake/exhaust valve clearance
(9)	Check/adjust the remote control operation
(10)	Adjust the propeller shaft alignment

(1) Check the injection timing

The fuel injection timing is adjusted so that engine performance may become the best condition. As for the inspection and adjustment of the fuel pump, it is based on the service manual of the ML pump. The fuel injection timing is adjusted by the following procedure.

1) Complete air bleeding from the fuel line and set the engine ready for starting.
2) See that the timing marks on fuel pump mounting flange and gear case are aligned.
3) Set the speed control lever at the operating position.
4) Disconnect the injection pipe on the fuel pump side for the No.1 cylinder. (Do not remove the delivery holder.)
5) Check the fuel discharge from the delivery holder while turning the crankshaft clockwise as seen from the radiator, and stop turning it at the same time when the fuel comes out. Wipe out the fuel of the delivery holder exit. Next, turn the crankshaft in the opposite direction (counterclockwise), and return it to about 20 degrees before top dead center.
6) Check again the fuel discharge from the delivery holder while turning the crankshaft clockwise, and stop turning it at the same time when the fuel comes out.
7) Read the timing scale on the flywheel from the hole on the flywheel housing. It is standard fuel injection timing if the timing mark position meets the fuel injection timing of the below table.

Model	Injection timing FID (FIC-Air) degree bTDC
3YM30	16 ± 1 (18 ± 1)
3YM20	22 ± 1 (24 ± 1)
2YM15	21 ± 1 (23 ± 1)

8) Repeat the step 5) to 7) a few times.

(Note)
Injection timing check for one cylinder is generally sufficient. If it is to be checked for all cylinders, check each cylinder in the ignition order of 1-3-2-1 (for 3 cyl. Eng.) or 1-2-1(for 2 cyl. Eng.).
The cylinder to be checked is not limited to the No.1 cylinder and any cylinder may be checked.

9) If the injection timing is out of the standard value, loosen the fuel pump mounting nut and incline the fuel injection pump toward or away from the engine for adjustment. Incline toward the engine to delay the timing, and away from the engine to advance it.

2. Inspection and adjustment

(2) Check the injection spray condition of a fuel nozzle

⚠ WARNING Wear protective glasses when testing injection from the fuel injection nozzle. Never approach the injection nozzle portion with a hand. The oil jetting out from the nozzle is at a high pressure to cause loss of sight or injury if coming into careless contact with it.

1) Injection pressure measurement

Standard fuel injection pressure

MPa (kgf/cm^2)
12.3-13.28 (125-135)

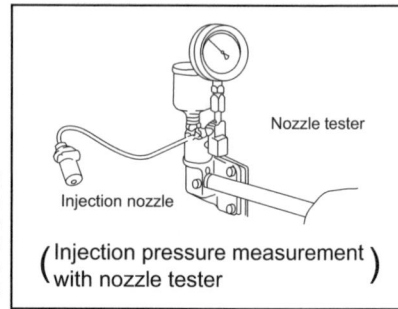

Nozzle tester

Injection nozzle

(Injection pressure measurement with nozzle tester)

[Notice]
As for the opening pressure of the brand-new fuel nozzle, about 0.5 MPa (5 kgf/cm) declines by the engine operation for about 5 hours because of the initial wear-out of a spring etc. Therefore, adjust 0.5 MPa (5 kgf/cm) higher than the standard value of the above table when adjusting a new fuel nozzle of a spare part.

Remove carbon deposit at the nozzle hole thoroughly before measurement.

a) Connect the fuel injection valve to the high pressure pipe of the nozzle tester.

b) Operate the nozzle tester lever slowly and read the pressure at the moment when the fuel injection from the nozzle starts.

c) If the measured injection pressure is lower than the standard level, replace the pressure adjusting shim with a thicker one.

Thickness of pressure adjusting shims mm	Injection pressure adjustment
0.1, 0.2, 0.3, 0.4, 0.5, 0.52, 0.54, 0.56, 0.58, 0.8	The injection pressure is increased by approx. 1.1 MPa (11 kgf / cm^2), when the adjusting shim thickness is increased by 0.l mm.

2. Inspection and adjustment

[Informative: Fuel injection valve structure]

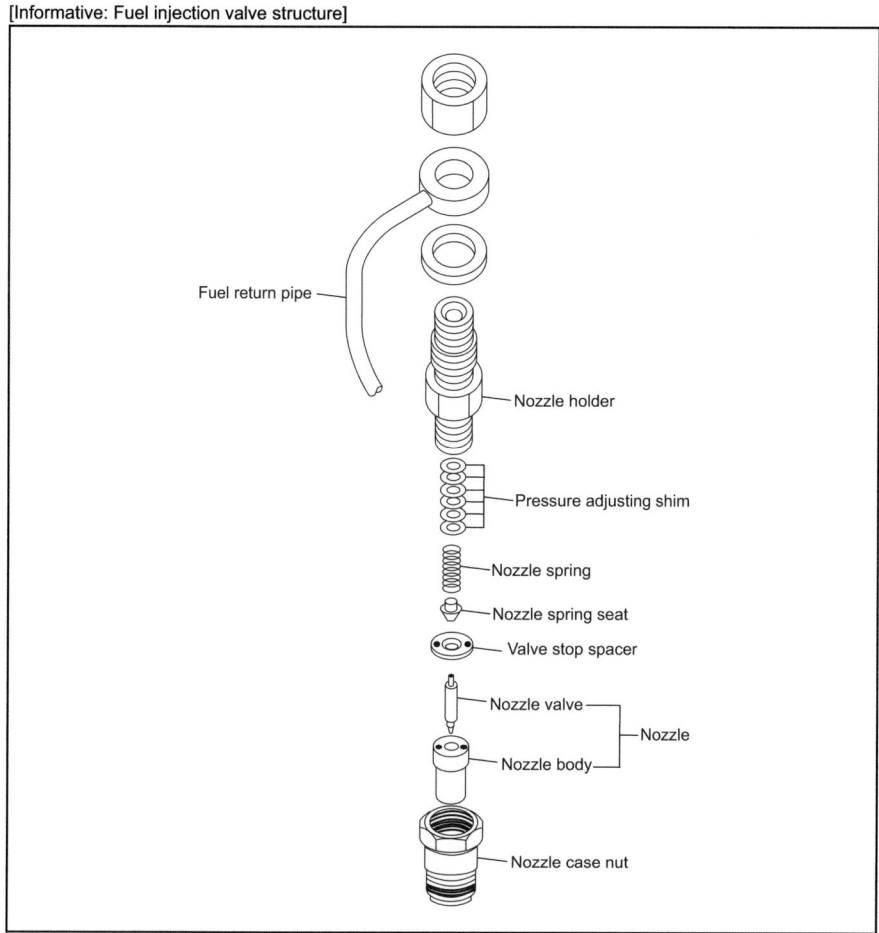

2. Inspection and adjustment

2) Spray pattern inspection
   After adjustment to the specified valve opening pressure, use a nozzle tester and check the spray pattern and seat oil-tightness.

   a) Seat oil tightness check
   - After injecting a few times, increase the pressure gradually. Hold the pressure for about 5 seconds at a little before the valve opening pressure of 1.96 MPa (20 kgf/cm^2), and check to see that oil does not drip from the tip end of the nozzle.
   - If extreme oil leak from the overflow joint exists during injection by the nozzle tester, check after retightening. If much oil is leaking, replace the nozzle assembly.

   b) Spray and injection states
   - Operate the nozzle tester lever at a rate of once or twice a second and check no abnormal injection.
   - If normal injection as shown below cannot be obtained, replace the fuel injection valve.
   - No extreme difference in angle(Θ)
   - Finely atomized spray
   - Excellent spray departure

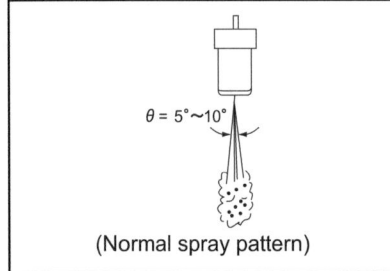
(Normal spray pattern)

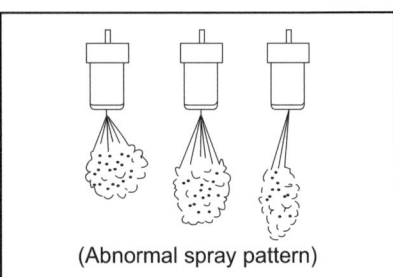
(Abnormal spray pattern)

3) Nozzle valve sliding test
   Wash the nozzle valve in clean fuel oil. Place the nozzle body vertically and insert the nozzle into the body to about 1/3 of its length. The valve is normal if it smoothly falls by its own weight into the body. In case of a new nozzle, remove the seal peel, and immerse it in clean diesel oil or the like to clean the inner and outer surfaces and to thoroughly remove rust-preventive oil before using the nozzle. Note that a new nozzle is coated with rust-preventive oil and is pasted with the seal peel to shut off outer air.

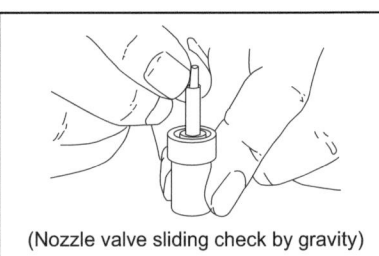
(Nozzle valve sliding check by gravity)

2. Inspection and adjustment

4) Nozzle punch mark

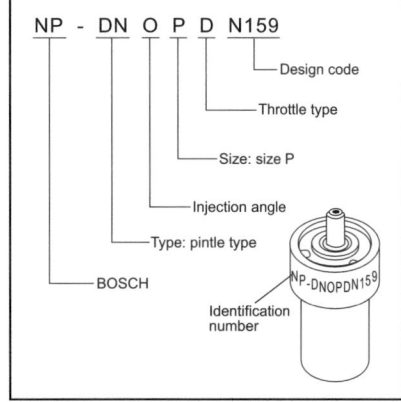

```
NP - DN O P D N159
              │ │ │ └─ Design code
              │ │ └─── Throttle type
              │ └───── Size: size P
              └─────── Injection angle
         └──────────── Type: pintle type
     └──────────────── BOSCH
                       Identification number
```

(3) Check the impeller of the cooling water pump (seawater pump)
The impeller must be replaced with new one periodically (every 1000 hrs or four years whichever comes first).
Refer to 2.2.5 (5) for the procedure.

(4) Clean & check the water passages
When it is used for a long time, cleaning of the cooling water passages is periodically necessary, because trash, scales, rust, and so on collect in the cooling water passages and the cooling performance declines.

(a) Cleaning and inspection of cooling water rubber hoses.
  1) Remove the cooling water rubber hoses for the fresh water line and the seawater line.
  2) Check the dust and trash inside the hoses. If necessary, clean it.
  3) Check the crack and the deterioration of the rubber hoses. If necessary, replace with new ones.
  4) Replace the used hose clips with new ones, if necessary.

(b) Heat exchanger inspection
  1) Cooler core inspection
     a) Inspect the inside of the tubes for rust or scale buildup from seawater, and clean with a wire brush if necessary.

NOTE: Disassemble and wash when the cooling water temperature reaches 85°C.

     b) Check the joints at both ends of the tubes for looseness or damage, and repair if loose, Replace if damaged or corroded
     c) Check tubes and replace if leaking.
     d) Clean any scale or rust off the outside of the tubes.
  2) Heat exchanger body inspection
     a) Check heat exchanger body and side cover for dirt and corrosion. Replace, if excessively corroded, or cracked.
     b) Inspect seawater and fresh water inlets and outlets, retighten any joints as necessary and clean the insides of the pipes.
     c) Check the exhaust gas intake flange and line, and replace if corroded or cracked.

2. Inspection and adjustment

3) Heat exchanger body water leakage test

   a) Compressed air/water tank test
      Fit rubber covers on the fresh water and seawater inlets and outlets. Place the heat exchanger in a water tank, feed in compressed air from the overflow pipe and check for any (water) leakage, (air bubbles).

Test pressure	0.20 MPa (2 kgf/cm^2)

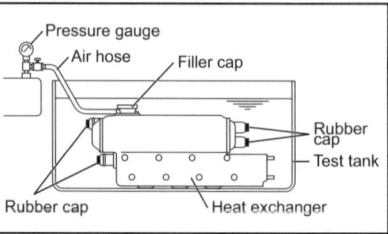

   b) Use of the tester
      Fit the fresh and seawater inlets and outlets with rubber covers and fill the fresh water tank with fresh water. Fit a pressure cap tester in place of the pressure cap, operate the pump for one minute and set the pressure at 0.15 Mpa (1.5 kgf/cm^2, 21.33 lb/in.^{2}). If there are any leaks the pressure will not rise. If there are no leaks the pressure will not fall.

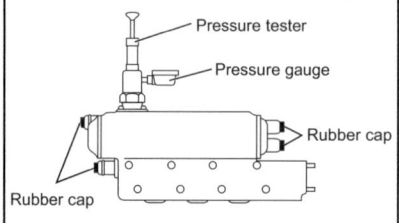

4) Pressure cap inspection

⚠ WARNING Do not open the pressure cap while the engine is running or right after stopping because high temperature steam will be blown out. Remove the cap only after the water cools down.

   a) Remove scale and rust and check the seat and seat valve, etc. for scratches or wear. Check the spring for corrosion or settling. Replace if necessary.

[NOTE]
Clean the pressure cap with fresh water as it will not close completely if it is dirty.

   b) Fit the adopter on the tester to the pressure cap. Pump until the pressure gauge is within the specified pressure range 0.074-0.103 MPa (0.75-1.05 kgf/cm^2) and note the gauge reading. The cap is normal if the pressure holds for six seconds. If the pressure does not rise, or drop immediately, inspect the cap and repair or replace as necessary.

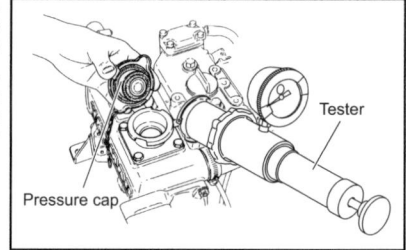

2. Inspection and adjustment

(5) Diaphragm assembly inspection
Inspect the diaphragm assy on the rocker arm cover.

1) Loosen screws, and remove a diaphragm assembly, and check whether oil and so on doesn't enter between the diaphragm and the cover. If oil and so on enters into the diaphragm assy, the diaphragm doesn't work well.
2) Check the damages of the diaphragm rubber and the spring. If necessary, replace with new ones.

[Notice]
1) When a diaphragm is damaged, pressure control inside the crankcase becomes insufficient, and troubles such as combustion defect and so on occur.

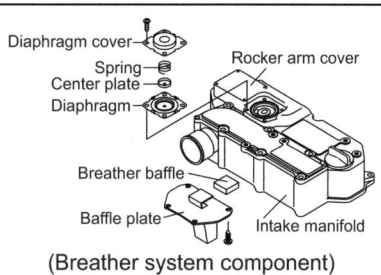

(Breather system component)

2) At lube oil replacement or lube oil supply, the amount of lube oil isn't to be beyond the standard upper limit. If the lube oil quantity is beyond the upper limit or an engine is operated beyond the allowable maximum angle of an engine, the amount of oil mist may be inducted in the combustion chamber and the oil hammer sometimes may occur.

(6) Adjust the tension of the alternator driving belt
Replace the alternator driving belt with new one every 1000 hours or four years, whichever comes first, even if there is no crack in the surface.
Refer to 2.2.2(4) for the procedure.

(7) Retighten all major nuts and bolts.
Retighten the major nuts and bolts bellow by the standard tightening torques. (See the table in 4.5.)

1) Cylinder head bolts
2) Crankshaft pulley bolts
3) Fuel injection nozzle set bolts
4) Fuel pump gear set bolts
5) Fuel injection line sleeve nut

Retighten the following major bolts as requested.

1) Rod bolts
2) Flywheel set bolts
3) Metal cap bolts

(8) Adjust intake/exhaust valve clearance.
Refer to 2.2.2(5) for the procedure.

(9) Check/adjust the remote control operation.
Refer to 2.2.2(6) for the procedure.

(10) Adjust the propeller shaft alignment.

1) The rubber tension of the flexible engine mounts is lost after many hours' use. This leads to a drop in vibration absorption performance, and also causes centering misalignment of the propeller shaft. (Refer to 5. "Flexible Engine Mount" in the installation manual for pleasure boat use.)

[Notice]
Be sure to replace the Yanmar flexible engine mounts every 1000 hours or 4 years, whichever comes first.

2) After replacing the flexible engine mounts, check unusual noise and vibration of the engine and the boat hull with increasing the engine speed gradually and lowering it.
3) If there is unusual noise and/or vibration, adjust the propeller shaft alignment. (Refer to 6.4.5 "Centering the Engine" in the installation manual for pleasure boat use.)

2. Inspection and adjustment

2.3 Adjusting the no-load maximum or minimum speed

1) After warming the engine up, gradually raise the speed and set it at the no-load maximum speed.
2) If the no-load maximum speed is out of the standard, adjust it by turning the high idle limiting bolt.
3) Then set the no-load minimum speed by adjusting the low idle limiting bolt.

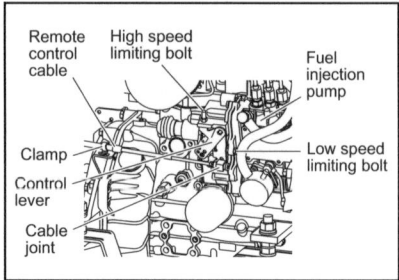

Standards (Unit : min $^{-1}$)

Model	No-load maximum speed	No-load minimum speed
3YM30	3850±25	
3YM20	3890±25	850±25
2YM15	3850±25	

2.4 Sensor Inspection

2.4.1 Oil pressure switch

Disconnect the connector from the oil pressure switch. Keep the voltameter probes in contact with the switch terminal and cylinder block while operating the engine. It is abnormal if circuit is closed.

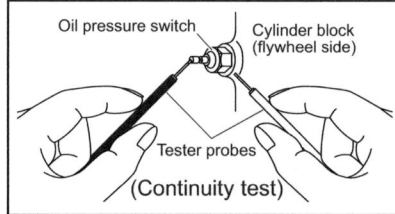

2.4.2 Thermo switch

Place the thermo switch in a container filled with antifreeze or oil. Heat it while measuring the fluid temperature. The switch is normal if the voltameter shows continuity when the fluid temperature is 93-97 deg C.

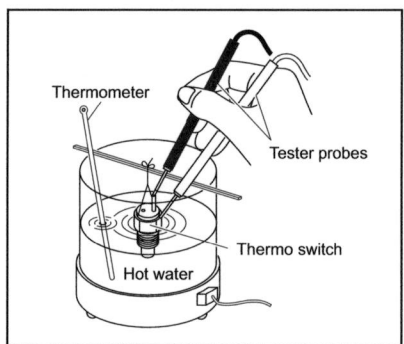

2.5 Thermostat inspection

(1) Put the thermostat in a beaker with fresh water, and heat it on an electric stove. The thermostat is functioning normally if it starts to open between 75-78 deg C, and opens until 8 mm or more at 90 deg C.
Replace the thermostat if it not functioning normally.

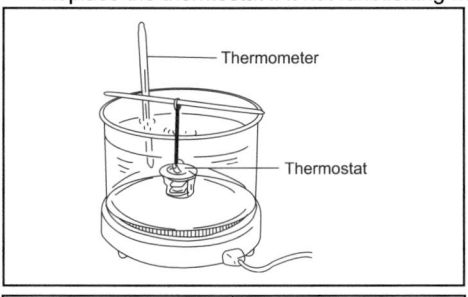

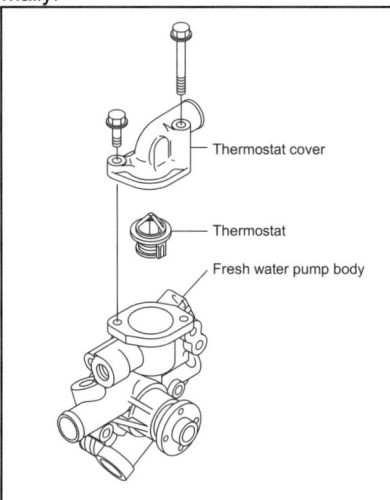

Valve opening Temperature (deg C)*	Full open lift (Temperature) (mm)
69.5-72.5	8 or more (85 deg C)

* Valve opening temperature is carved on the flange.

(2) Normally, the thermostat should be inspected every 500 hours or operation, but, it should be inspected before this if the cooling temperature rises abnormally or white smoke is emitted for a long time after engine starting.

(3) Replace the thermostat every year or 2000 hours of operation (whichever comes first).

2.6 Adjusting operation

Perform the adjusting operation for an engine as follows after the maintenance job:

2.6.1 Preliminary precautions

Before making a test run, make sure of the following points.

(1) Warm the engine up.

(2) Remove any precipitation from the F.O. filter. water separator, and F.O. tank.

(3) Use only lube oil recommended by Yanmar.

(4) Be sure to add Long Life Coolant Antifreeze (LLC) to cooling fresh water.

(5) Provide good ventilation in the engine room.

2.6.2 Adjusting operation procedure

1) Supply the fuel oil, lubricating oil and cooling water.

Note:
Check the levels of the lube oil and cooling water again after test running (for about 10 minutes) and add as required.

2) Start the engine, and carry out idling at a low speed (700 to 900 min^{-1}) for a few minutes.
3) Run in the engine for about five minutes at the rated speed (no-load). Check any water, fuel or lube oil leakage and existence of abnormal vibration or noise. Also check the oil pressure, cooling water temperature and exhaust gas color.
4) Adjust the no-load minimum and maximum speed. (Refer to 2.3.)
5) Perform loaded operation as required.

2. Inspection and adjustment

2.6.3 Check points and precautions during running

Step	Item	Instructions	Precautions
1	Checks before operation	1) Make sure that the sea cock is open. 2) Make sure there is enough lube oil and (fresh) cooling water. 3) Operate the remote control handle and check if the device connected to the engine works property.	3) Lamp should go off when engine is running.
2	No load operation; warm up operation	1) When the lube oil temperature is raised to allow the engine to start, the pilot lamp goes off. 2) When the engine is started, check the following: • There is no leakage of water, fuel and lube oil. • Exhaust gas does not leak when the engine is started. • here is no abnormal indication on the instrument panel. • There is no abnormality in cooling water discharge, engine vibration. or engine sound. 3) To warm up the engine, operate at low speed for about 5 minutes, then raise the speed to the rated speed and then to max. speed.	2) • Fit leaks if any. • Check the intake/ exhaust valves, fuel injection nozzle and cylinder head. 3) Do not raise the engine speed abruptly.
3	Cruising (load) operation	1) Do not operate the engine at full load yet, but raise the speed gradually for about 10 minutes until it reaches the rated speed. 2) Make sure that exhaust gas color and temperature are normal. 3) Check the instrument panel and see if the water temperature and oil pressure are normal.	
4	Stopping the engine	1) Before stopping the engine, operate it at 650-700 min^{-1} for about 5 minutes. 2) Raise engine speed to 1,800 min^{-1} jut before stopping the engine and idle the engine for about 3-4 seconds.	1) Stopping the engine suddenly during high speed operation increases the temperature of engine parts. 2) This procedure prevents carbon from being deposited on the valve seats, etc.
5	Checks after stopping the engine	1) Check again for water and oil leaks. 2) Make sure that no nuts and bolts are loose. 3) Close the sea cock and fuel cocks. 4) When the temperature is expected to fall below freezing, drain the seawater. 5) Turn off the battery switch.	1) Check the oil seal area. 2) Especially the engine installation bolts. 4) Drain from the seawater pump.

2.7 Long storage

Observe the following instructions when the engine is to be stored for a long period without operation:

1) Do not drain cooling water in the cold season or before the long storage.

[NOTICE]
Negligence of adding anti-freeze will cause the cooling water remaining inside the engine to be frozen and expanded to damage the engine parts.

2) Remove the mud, dust and oil deposit and clean the outside.
3) Perform the nearest periodic inspection before the storage.
4) Drain or fill the fuel oil fully to prevent condensation in the fuel tank.
5) Disconnect the battery cable from the battery negative (-) terminal.
6) Cover the silencer, air cleaner and electric parts with PVC (Poly Vinyl Chloride) cover to prevent water and dust from depositing or entrance.
7) Select a well-ventilated location without moisture and dust for storage.
8) Perform recharging once a month during storage to compensate for self-discharge.
9) When storing an engine for long time, run the engine periodically according to the following procedure because the rust occurrence inside the engine, the rack agglutination of the fuel pump, and so on are likely to occur. (In case that the engine is equipped with a boat.)

   a) Replace the lube oil and the filter before the engine running.
   b) Supply fuel if the fuel in the fuel tank was removed, and bleed the fuel system.
   c) Confirm that there is the coolant in the engine.
   d) Operate the engine at the low idling speed for about five minutes. (If it can be done, once a month)

3. Troubleshooting

3.1 Preparation before troubleshooting

If the signs of a trouble appear, it is important to lecture on the countermeasure and treatment before becoming a big accident not to shorten the engine life.

When the signs of a trouble appear in the engine or a trouble occurs, grasp the trouble conditions fully by the next point and find out the cause of sincerity according to the troubleshooting. Then repair the trouble, and prevent the recurrence of the trouble.

1) What's the occurrence phenomenon or the trouble situation?
   (e.g. Poor exhaust color)
2) Investigation of the past records of the engine
   Check a client control ledger, and examine the history of the engine.
   - Investigate the engine model name and the engine number. (Mentioned in the engine label.)
     Examine the machine unit name and its number in the same way.
   - When was the engine maintained last time?
   - How much period and/or time has it been used after it was maintained last time?
   - What kind of problem was there on the engine last time, and what kind of maintenance was dane?
3) Hear the occurrence phenomenon from the operator of the engine in detail.
   5W1H of the occurrence phenomenon : the investigation of when (when), where (where),
                            who (who), what (what), why (why) and how (how)
   - When did the trouble happen at what kind of time?
   - Was there anything changed before the trouble?
   - Did the trouble occur suddenly, or was there what or a sign?
   - Was there any related phenomenon.
     (e.g. Poor exhaust color and starting failure at the same time)
4) After presuming a probable cause based on the above investigation, investigate a cause systematically by the next troubleshooting guide, and find out the cause of sincerity.

3.2 Quick reference chart for troubleshooting

It is important to thoroughly under stand each system and the function of all of the parts of terse systems. A careful study of the engine mechanism will make this possible. When problems arise, it is important to carefully observe and analyze the indications of trouble in order to save time in determining their cause. Begin by checking the most easily identifiable causes of difficulty. Where the cause of the difficulty is not readily apparent, make a thorough examination of the system from the very beginning, proceeding until the point of trouble can be determined. While experience is an important factor in pinpointing engine problems, careful study and understanding of the engine mechanism combined with good common sense will help you to rapidly become more expert at troubleshooting.

3. Troubleshooting

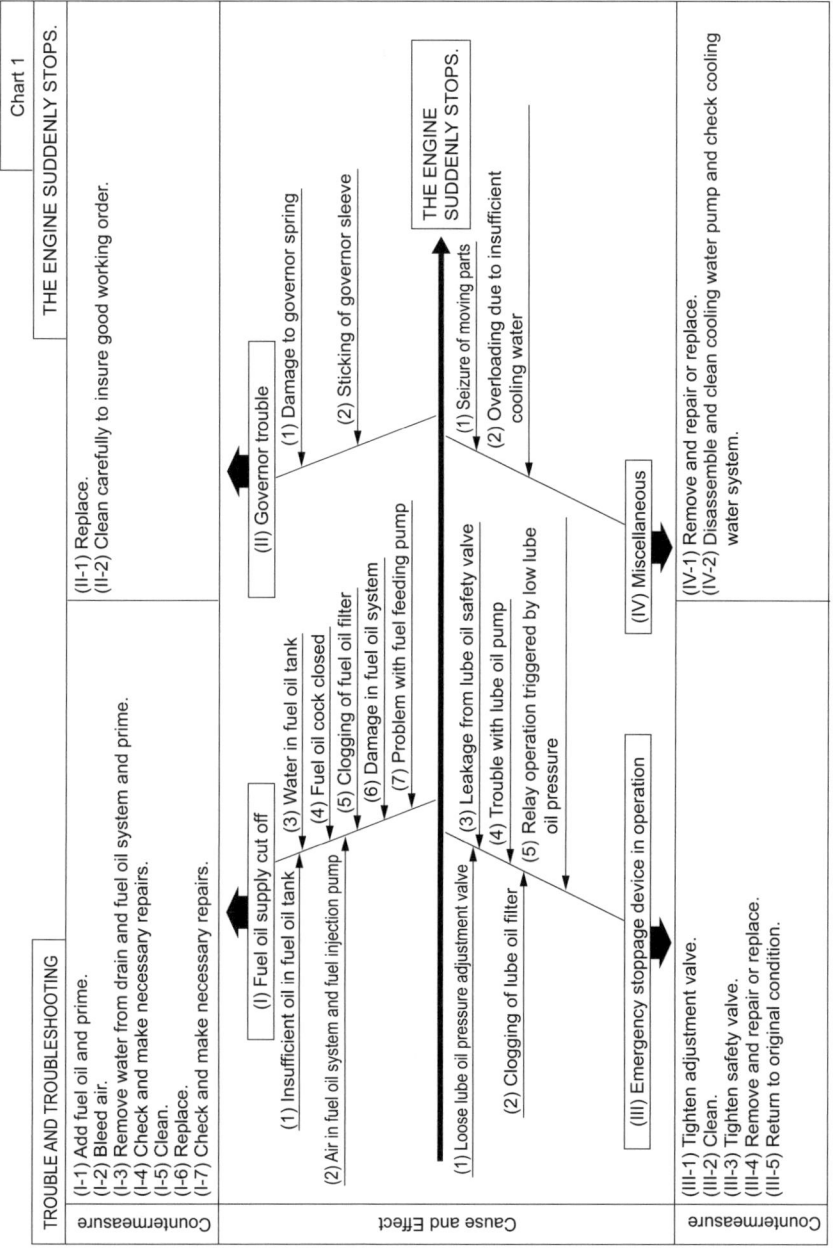

3. Troubleshooting

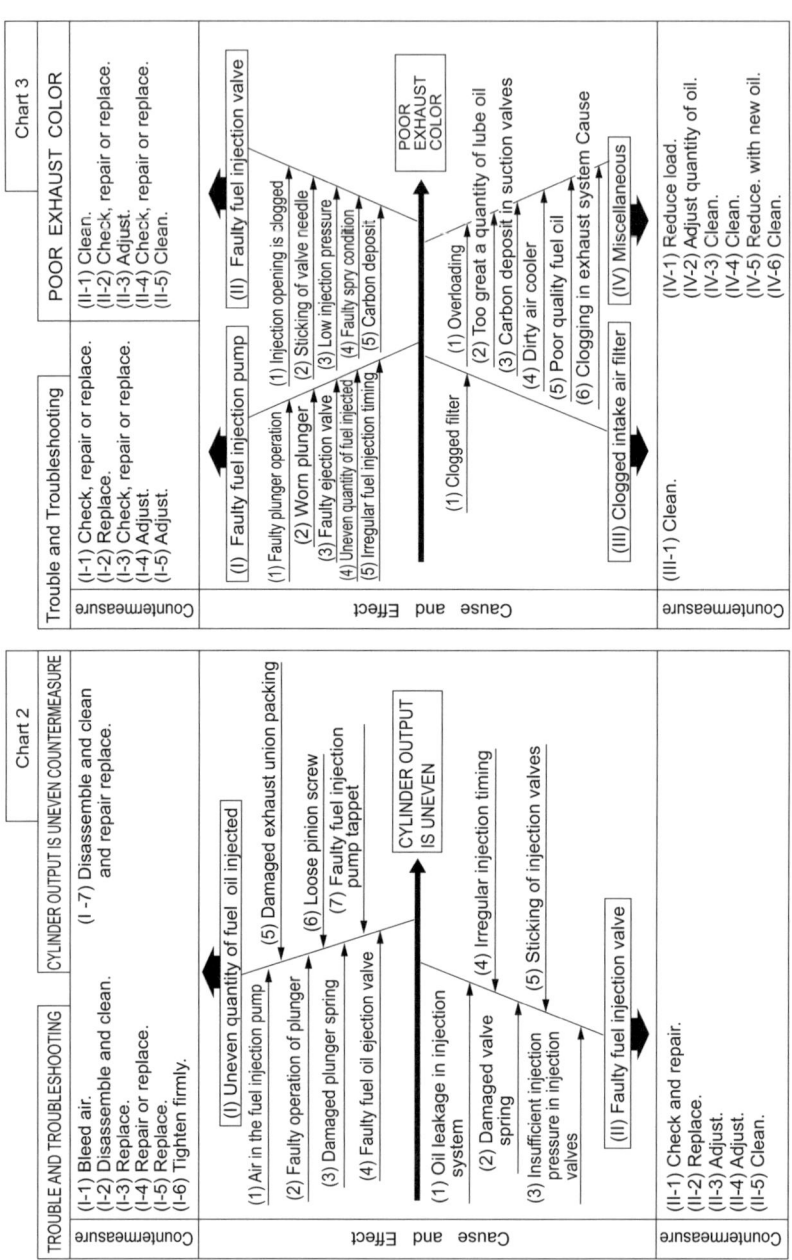

3. Troubleshooting

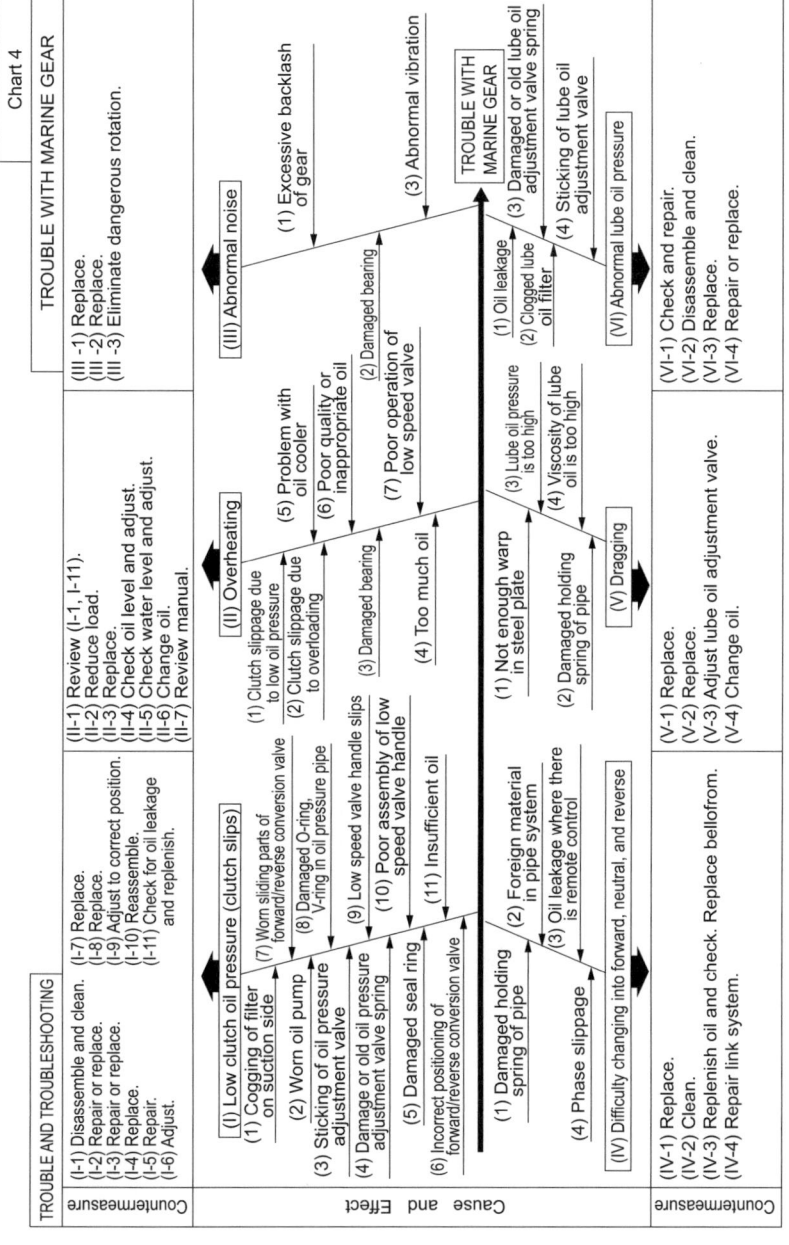

Chart 4

3. Troubleshooting

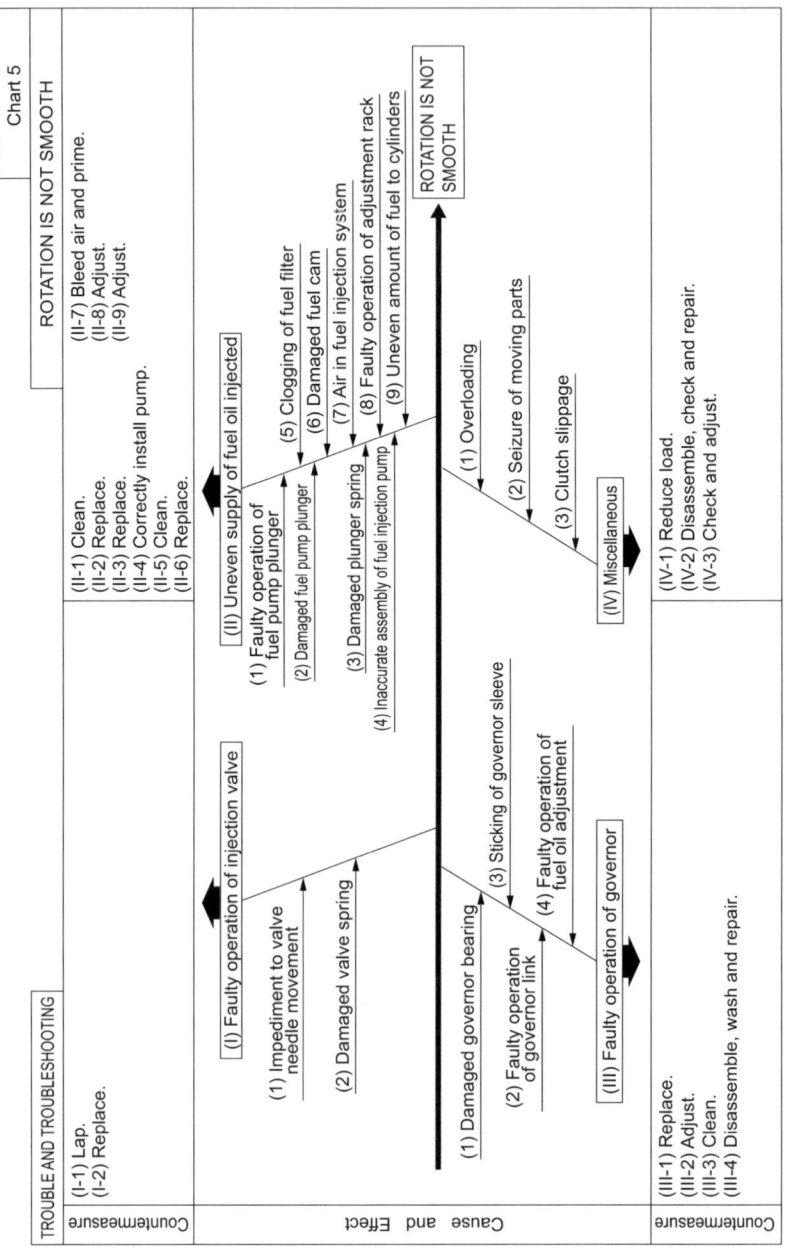

Chart 5

3. Troubleshooting

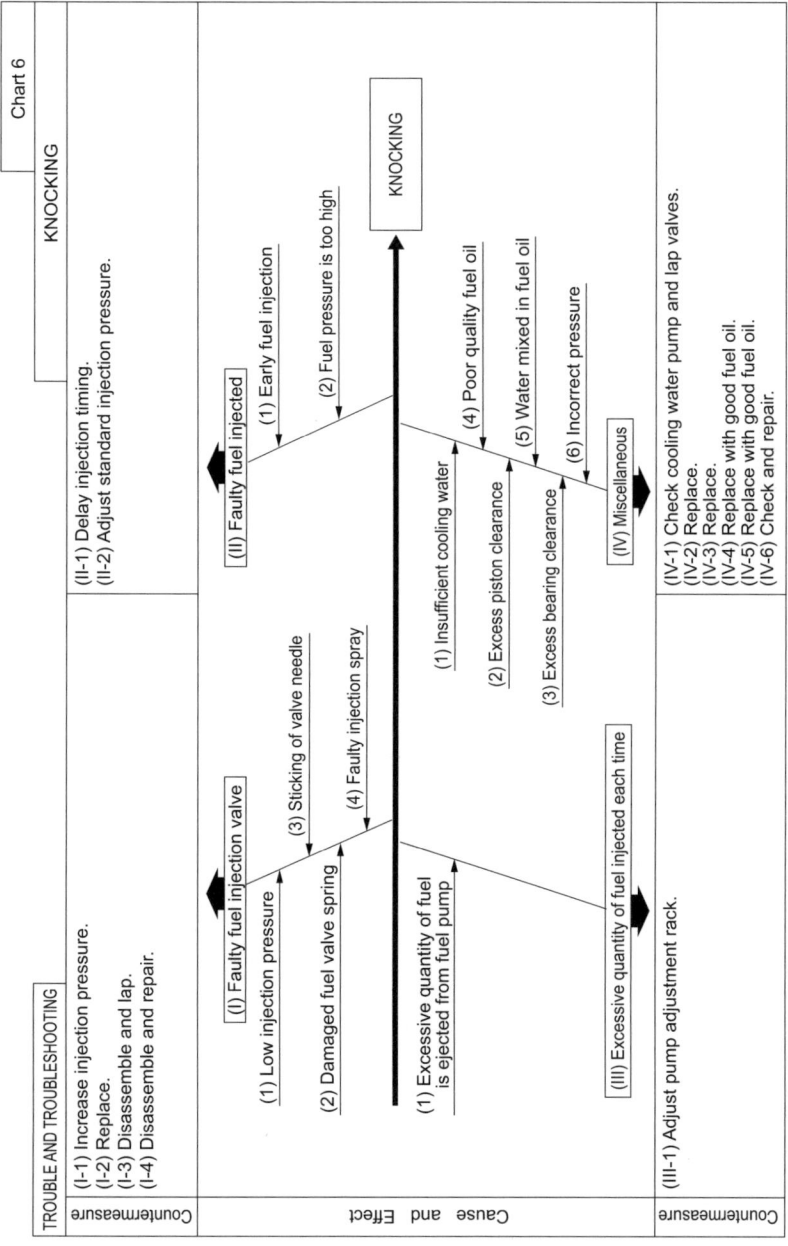

Chart 6

3. Troubleshooting

Chart 7 — TROUBLE WITH STARTING

TROUBLE AND TROUBLESHOOTING

Cause and Effect

(I) Pinion gears do not mesh
- (1) Loose battery engage magnet terminal
- (2) Faulty connection in starting switch
- (3) Cut battery engage magnet coil
- (4) Rough cap movement
- (5) Edges of gear teeth misshapen
- (6) Faulty positioning of pinion and ring gears
- (7) Seizure of starter metal

(II) Ring gears do not mesh
- (1) Loose battery starter terminal
- (2) Faulty connection in engage magnet switch
- (3) Worn brushes
- (4) Commutator rough and dirty
- (5) Cut starter coil
- (6) Worn commutator
- (7) Slippage of starter clutch
- (8) Excessive resistance between battery and starter
- (9) Insufficient battery charge

(III) Fuel oil not injected
- (1) Imperfect priming of fuel oil system
- (2) Injection cut off due to faulty governor
- (3) Clogging of fuel inlet filter
- (4) Insufficient fuel oil in tank
- (5) Fuel oil tank cock closed
- (6) Clogging in fuel oil pipe
- (7) Damaged fuel oil supply pump
- (8) Water in fuel oil tank

(IV) Faulty fuel injection valve
- (1) Faulty valve seat
- (2) Sticking of valve needle
- (3) Worn valve needle
- (4) Clogged outlet
- (5) Low injection pressure

STARTING TROUBLE

(V) Faulty fuel injection pump
- (1) Worn plunger
- (2) Damaged plunger spring
- (3) Plunger sticks
- (4) Oil leakage from exhaust valve
- (5) Air inside pump
- (6) Damaged exhaust valve spring

(VI) Faulty fuel injection system
- (1) Fuel injection pump timing irregular
- (2) Loose high pressure fuel pipe
- (3) Damaged high pressure fuel pipe
- (4) Air in high pressure fuel pipe

(VII) Leakage of pressurized air
- (1) Air leakage from suction/exhaust valves
- (2) No tappet clearance
- (3) Faulty gasket packing
- (4) Upper part of cylinder liner worn
- (5) Worn piston ring
- (6) Piston ring sticks
- (7) Insufficient tightening of head bolts
- (8) Damaged valve spring

(VIII) Miscellaneous
- (1) Incorrect thickness of gasket packing
- (2) Faulty installation of governor ring and lever
- (3) Governor handle is in stop position
- (4) Faulty engine starter
- (5) Clogging of suction/exhaust pipes

Countermeasure

(I)
- (I-1) Tighten.
- (I-2) Repair using sandpaper or replace.
- (I-3) Replace.
- (I-4) Repair using sandpaper and then grease.
- (I-5) Adjust.
- (I-6) Adjust.
- (I-7) Replace.

(II)
- (II-1) Tighten.
- (II-2) Repair using sandpaper.
- (II-3) Replace.
- (II-4) Repair using sandpaper (Type 500~600).
- (II-5) Replace.
- (II-6) Undercut and repair or replace.
- (II-7) Replace.
- (II-8) Replace with thicker or shorter wire.
- (II-9) Charge.

(III)
- (III-1) Prime well.
- (III-2) Adjust.
- (III-3) Clean out matter causing clogging.
- (III-4) Add fuel to fuel tank.
- (III-5) Open cock.
- (III-6) Clean.
- (III-7) Disassemble and repair or replace.
- (III-8) Drain water from fuel system and prime.

(IV)
- (IV-1) Lap.
- (IV-2) Lap.
- (IV-3) Replace.
- (IV-4) Clean or replace.
- (IV-5) Adjust.

(V)
- (V-1) Replace plunger and barrel as a unit.
- (V-2) Replace.
- (V-3) Disassemble and repair or replace.
- (V-4) Lap valves.
- (V-5) Bleed air.
- (V-6) Replace.

(VI)
- (VI-1) Adjust.
- (VI-2) Tighten firmly.
- (VI-3) Replace.
- (VI-4) Bleed air.

(VII)
- (VII-1) Lap valves.
- (VII-2) Adjust.
- (VII-3) (VII-4) (VII-5) Replace.
- (VII-6) Disassemble and repair or replace.
- (VII-7) Tighten tightening nuts uniformly.
- (VII-8) Replace.

(VIII)
- (VIII-1) Replace.
- (VIII-2) Adjust.
- (VIII-3) Move governor handle to acceleration position.
- (VIII-4) Check and repair.
- (VIII-5) Clean.

3. Troubleshooting

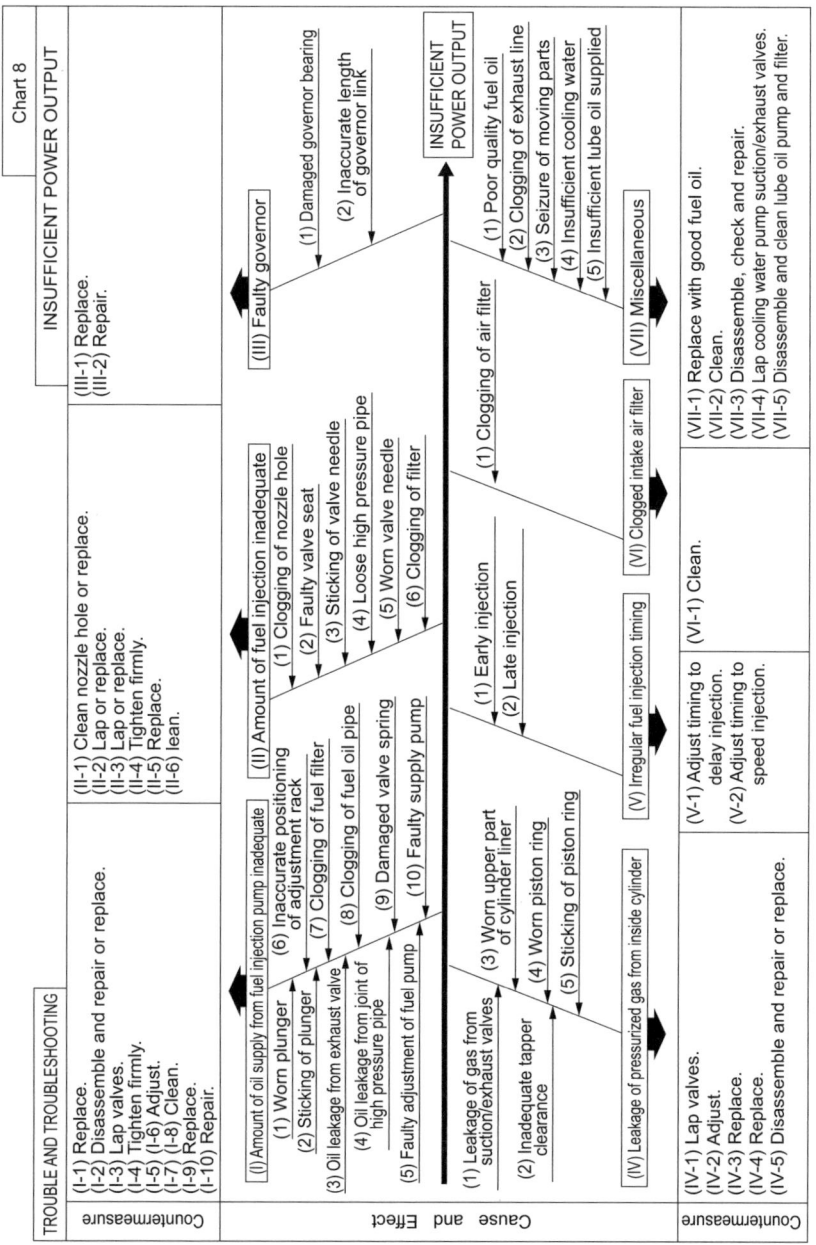

Chart 8

3. Troubleshooting

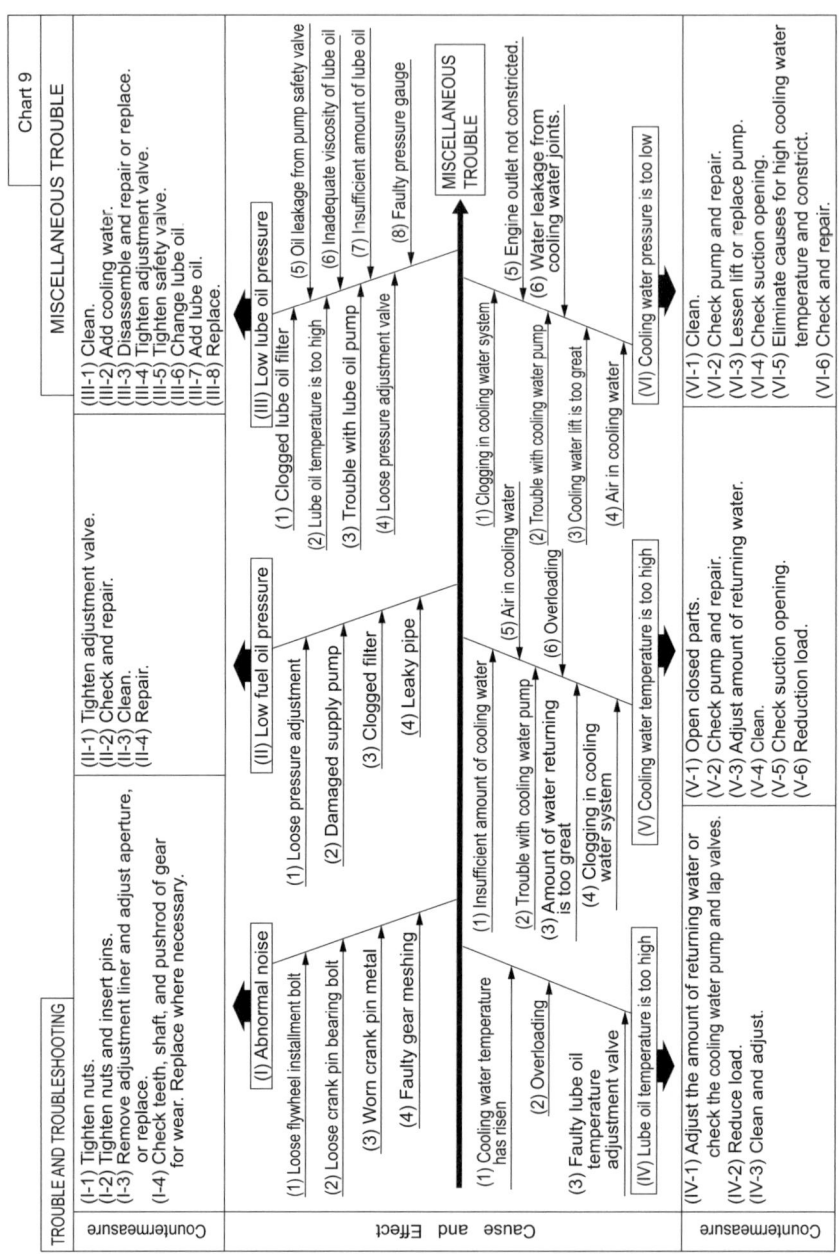

Chart 9

3.3 Troubleshooting (Concerning engine and fuel injection equipment)

Malfunctions	Causes	Remedies
The engine does not operate 1. Fuel oil is not injected from the injection pump	1. There is no fuel oil in the fuel tank	Supply fuel and bleed the system
	2. The fuel line from the fuel tank is blocked	Clean or replace
	3. The fuel is clogged	Clean or replace
	4. There is air in the fuel filter or the pump chamber	Bleed the system
	5. The accelerator linkage is not properly connected	Repair
	6. The magnet valve wiring is broken or its armature is sticking	Repair or replace
	7. The feed pump blades are sticking, and therefore not operating	Repair or replace
	8. The drive gear or woodruff key is broken	Replace
2. Injection timing is incorrect	1. The drive gear or belt connections are incorrect	Repair
	2. The injection pump is incorrectly installed on the engine	Repair and adjust injection timing
	3. The roller holder assembly's roller or pin is worn excessively	Replace the assembly
	4. The plunger is worn excessively	Replace the distributor assembly
3. The nozzle does not operate	1. The nozzle or nozzle holder is functioning incorrectly	Inspect, then repair or replace
4. The engine operates but only for short time	1. The pipe(s) to the injection pump is blocked or the fuel filter is clogged	Clean or replace the pipe(s) or fuel filter
	2. The fuel oil contains air or water	Bleed of air or replace the fuel oil
	3. The feed pump's delivery quantity (or pressure) is insufficient	Repair or replace
5. The engine "knocks"	1. The injection timing is too advanced	Readjust the timing
	2. The nozzle or nozzle holder is functioning incorrectly	Inspect, then repair or replace

3. Troubleshooting

Malfunctions	Causes	Remedies
The engine exhaust contains smoke and the engine "knocks"	1. The injection timing is incorrect 2. The nozzle or nozzle holder is functioning incorrectly 3. The injection quantity is excessive	Readjust the timing Inspect, then repair or replace Readjust
The engine output is unstable	1. The fuel filter element is clogged and fuel oil delivery is poor 2. The amount of fuel or pressure delivered by the feed pump is too little 3. The injection pump is sucking air 4. The regulating valve is stuck in the open position 5. The plunger is sticking and does not travel its full stroke 6. The plunger spring is broken 7. The control sleeve is not sliding smoothly 8. The governor lever is not operating properly or is worn excessively 9. The delivery valve spring is broken 10. The delivery valve is not sliding properly 11. The nozzle or the nozzle holder is not functioning properly 12. The injection timing is incorrect	Clean or replace Inspect and repair Inspect and repair Replace Replace the distributor assembly Replace Repair or replace Repair or replace Replace Repair or replace Inspect, and then repair or replace Readjust
Insufficient output 1. The injection quantity is insufficient	1. The specified full-load injection quantity is not delivered 2. The control lever is not reaching the maximum speed position 3. The governor spring is weak and therefore the governed speed is too low 4. The plunger is worn 5. The delivery valve seating portions are damaged	Readjust Readjust Replace Replace the distributor assembly Replace
2. The injection timing is too advanced and the engine is "knocking"		Readjust
3. The injection timing is too retarded and the engine is overheating or the exhaust contains smoke		Readjust

3. Troubleshooting

Malfunctions	Causes	Remedies
4. The nozzle or the nozzle holder is not functioning properly		Inspect and then repair or replace
The engine cannot reach its maximum speed	1. The governor spring is too weak or is improperly adjusted	Readjust or replace
	2. The control lever is not reaching the maximum-speed position	Readjust
	3. The nozzle's injection operation is poor	Repair or replace
The engine's maximum speed is too high	1. The governor spring is too strong or is improperly adjust	Readjust or replace
	2. The governor flyweights or governor sleeve movement is not smooth	Repair or replace
Idling is unstable	1. The injection quantities are not uniform (the delivery valve is not operating properly)	Inspect or replace
	2. The governor's idling adjustment is improperly adjusted	Readjust
	3. The plunger is worn	Replace the distributor assembly
	4. The plunger spring is broken	Replace
	5. The rubber damper is worn.	Replace
	6. The governor lever shaft pin is worn excessively	Replace
	7. The feed pump blades are not operating properly	Repair or replace
	8. The regulating valve is stuck in the open position	Replace
	9. The fuel filter element is clogged and therefore fuel oil delivery is poor	Clean or replace
	10. The nozzle or the nozzle holder is not functioning properly	Inspect and then repair or replace

3.4 Troubleshooting by measuring compression pressure

Compression pressure drop is one of major causes of increasing blowby gas (lubricating oil contamination or increased lubricating oil consumption as a resultant phenomenon) or starting failure. The compression pressure is affected by the following factors:

1) Degree of clearance between piston and cylinder
2) Degree of clearance at intake/exhaust valve seat
3) Gas leak from nozzle gasket or cylinder head gasket

In other words, the pressure drops due to increased parts wear and reduced durability resulting from long use of the engine.

A pressure drop may also be caused by scratched cylinder or piston by dust entrance from the dirty air cleaner element or worn or broken piston ring. Measure the compression pressure to diagnose presence of any abnormality in the engine.

(1) Compression pressure measurement method

1) After warming up the engine, remove the fuel injection pipe and valves from the cylinder to be measured.
2) Crank the engine before installing the compression gage adapter.
   a) Perform cranking with the stop handle at the stop position (no injection state).
   b) See 4.2.3(2) in Chapter 4 for the compression gage and compression gage adapter.
3) Install the compression gage and compression gage adapter at the cylinder to be measured.
   a) Never forget to install a gasket at the tip end of the adapter.
4) With the engine set to the same state as in 2)a), crank the engine by the starter motor until the compression gage reading is stabilized.

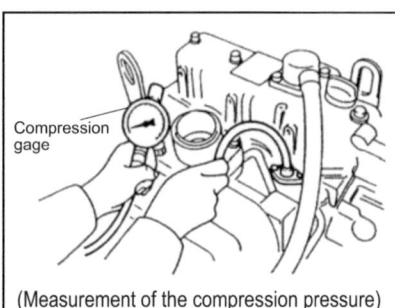

Compression gage

(Measurement of the compression pressure)

(2) Standard compression pressure

Engine compression pressure list (reference value)

Model	Compression pressure at 250 min^{-1} MPa (kgf/cm^2)		Deviation among cylinders MPa (kgf/cm^2)
	Standard	Limit	
3YM30	3.43 ± 0.1 (35 ± 1)	2.75 ± 0.1 (28 ± 1)	0.2 ± 0.3 (2 ± 3)
3YM20/2YM15	3.23 ± 0.1 (33 ± 1)	2.55 ± 0.1 (26 ± 1)	0.2 ± 0.3 (2 ± 3)

3. Troubleshooting

(3) Engine speed and compression pressure (for reference)

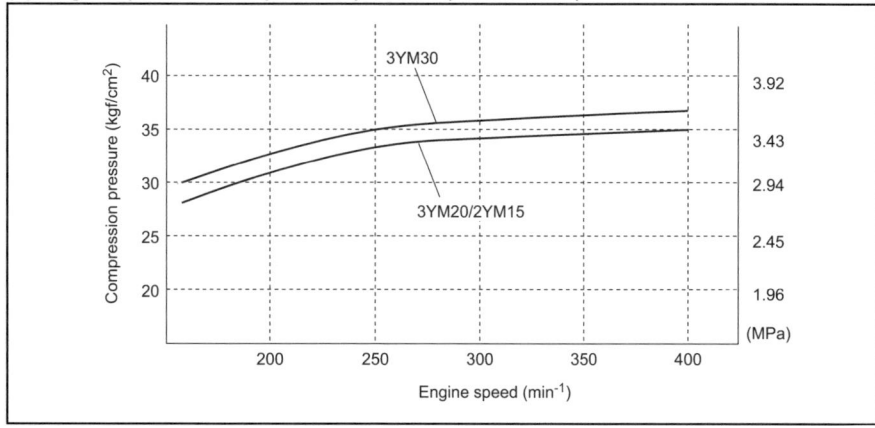

(4) Measured value and troubleshooting
When the measured compression pressure is below the limit value, inspect each part by referring to the table below.

No.	Item	Cause	Corrective action
1	• Air cleaner element	• Clogged element • Broken element • Defect at element seal portion	• Clean the element. • Replace the element.
2	• Valve clearance	• Excessive or no clearance	• Adjust the valve clearance. (See 2.2.2(5) in Chapter 2.)
3	• Valve timing	• Incorrect valve clearance	• Adjust the valve clearance. (See in Chapter2.)
4	• Cylinder head gasket	• Gas leak from gasket	• Replace the gasket. • Retighten the cylinder head bolts to the specified torque. (See 5.2.5 in Chapter 5.)
5	• Intake / exhaust vale • Valve seat	• Gas leak due to worn valve seat or foreign matter trapping • Sticking valve	• Lap the valve seat. (See 5.2.2 in Chapter 5.) • Replace the intake/exhaust valve.
6	• Piston • Piston ring • Cylinder	• Gas leak due to scratching or wear	• Perform honing and use an oversized part.

4. Disassembly and reassembly

4.1 Disassembly and reassembly precautions

(1) Disassembly
- Take sufficient time to accurately pin-point the cause of the trouble, and disassemble only those parts which are necessary.
- Be careful to keep all disassembled parts in order.
- Prepare disassembly tools.
- Prepare a cleaner and cleaning can.
- Clear an adequate area for parts and prepare a container(s)
- Drain cooling water (seawater, fresh water) and lube oil.
- Close the seacock

(2) Reassembly
- Sufficiently clean and inspect all parts to be assembled.
- Coat sliding and rotating parts with new engine oil when assembling.
- Replace all gaskets and O-rings.
- Use a liquid packing agent as necessary to prevent oil/water leaks.
- Check the oil and thrust clearances. etc. of parts when assembling
- Make sure you use the correct bolt/nut/washer.
- Tighten main bolts/nuts to the specified torque. Be especially careful not to over tighten the aluminum alloy part mounting bolts.
- Align match marks (if any) when assembling. Make sure that the correct sets of parts are used for bearings, pistons, and other parts where required.

4. Disassembly and reassembly

4.2 Disassembly and reassembly tools

The following tools are required when disassembling and reassembling the engine.
Please use them as instructed.

4.2.1 General hand tools

Name of tool	Illustration	Remarks
Wrench		Size : 10 x 13
Wrench		Size : 12 x 14
Wrench		Size : 17 x 19
Wrench		Size : 22 x 24
Screwdriver		
Steel hammer		Local supply

4. Disassembly and reassembly

Name of tool	Illustration	Remarks
Copper hammer		Local supply
Mallet		Local supply
Nippers		Local supply
Pliers		Local supply
Offset wrench		Local supply 1 set
Box spanner		Local supply 1 set
Scraper		Local supply

4. Disassembly and reassembly

Name of tool	Illustration	Remarks
Lead rod		Local supply
File		Local supply 1 set

Rod spanner for hexagon socket head screws		Local supply
		Size : 6 mm
		        8 mm
		        10 mm

Stariing pliers
  Hole type
  Shaft type
	S - 0
	
	H4 - H8
	
	S = Hole type
	H = Shaft type
		Local supply

4. Disassembly and reassembly

4.2.2 Special hand tools

Name of tool	Illustration	Remarks
Piston pin insertion/ extraction tool	Dimensions: 20, 80, 12, 25	(Piston pin extractor) (Extraction of piston pin) (Insertion of piston pin)
Connecting rod small end bushing insertion/extraction tool	Dimensions: 20, 80, $22_{-0.6}^{-0.3}$, $24_{-0.6}^{-0.3}$ * Locally manufactured	(Extraction)
Intake and exhaust valve insertion/ extraction tool	Dimensions: 20, 75, 5.5, 9.5 * Locally manufactured	
Crankshaft pulley insertion tool	Dimensions: 50, 26, 6, $\phi 20$, $\phi 27_{-0.2}^{-0.1}$, M12 x P1.25	

4. Disassembly and reassembly

Name of tool	Illustration	Remarks
Lube oil filter case remover		
Piston ring replacer (for removal / installation of piston ring)		
Valve lapping tool (Rubber cap type)		
Valve lapping powder	Code No. 28210-000070	
Feeler gauge		
Pulley puller	(Local supply)	(Removing the coupling)

4. Disassembly and reassembly

Name of tool	Illustration	Remarks
Press tool 1 for filler neck	(Tool 1)	-
Press tool 2 for filler neck copper tube	(Tool 2)	-
Stem seal insertion (for inserting stem seal)	mm | d1 | d2 | d3 | L1 | L2 | L3 | | 15.0 | 21 | 12 | 11.0 | 65 | 4 or more |	-

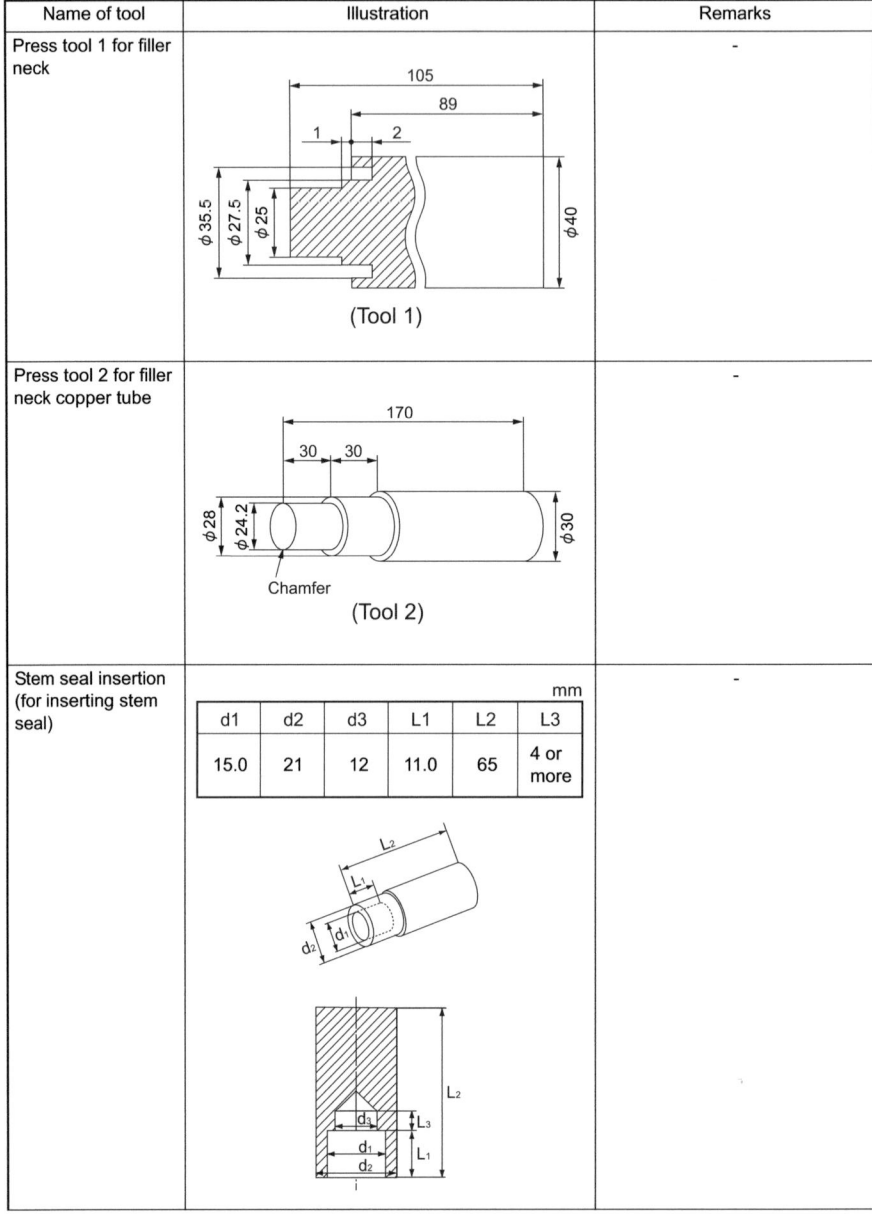

4.2.3 Measuring instruments
(1) Application of tools

Name of tool	Illustration	Remarks
Vernier calipers		0.05 mm 0-150 mm
Micrometer		0.01 mm 0-25 mm 25-50 mm 50-75 mm 75-100 mm 100-125 mm 125-150 mm
Cylinder gauge		0.01 mm 18-35 mm 35-60 mm 50-100 mm
Thickness gauge		0.05-2 mm
Torque wrench		128 N·m (0-13 kgf·m)
Nozzle tester		0-49 Mpa (0-500 kgf/cm^2)

4. Disassembly and reassembly

(2) Use of tools

No.	Name of tool	Use	Illustration
1	Dial gauge	Measures shaft bending, distortions of levelness, and gaps.	
2	Test indicator	Measures narrow and deep places, which cannot be measured with dial gauge.	
3	Magnetic stand	Keeps the dial gauge firmly in position, thereby permitting it to be used at various angles.	
4	Micrometer	Measures the outer diameter of the crank shaft, piston, piston pin, etc.	
5	Cylinder gauge	Measure the inner diameter of the cylinder liner and rod metal.	
6	Vernier calipers	Measures various outer diameters, thicknesses, and widths.	
7	Depth micrometer	Measures sinking of valves.	
8	Square	Measures distortion in position of springs and perpendicularity of parts.	

72

4. Disassembly and reassembly

No.	Name of tool	Use	Illustration
9	V Block	Measures shaft distortion.	
10	Torque wrench	Used to tighten bolts and nuts to standard torque.	
11	Thickness gauge	Measures the distance between the ring and ring groove, and between the shaft and shaft joint at time of assembling.	
12	Cap tester	Check for leakage in the fresh water system.	
13	Battery current tester	Checks density of antifreeze and charging condition of battery fluid.	
14	Nozzle tester	Checks the shape and pressure of spray emitted from the fuel injection valve at the time of injection.	
15	Digital thermostat	Measures temperature of various parts.	Float

4. Disassembly and reassembly

No.	Name of tool		Use	Illustration
16	Rotation gauge	Contact type	Measures rotation speed by placing at the indentation hole of the revolving shaft.	
		Photoelectric type	Measures rotation speed by using a reflector seal which is placed on the exterior of the revolving shaft	
		High pressure fuel pipe clamp type	Measures rotation speed without reference to revolving shaft center or the exterior of the revolving shaft.	
17	Circuit tester		Measures the resistance, voltage, and continuity of the electric circuit.	
18	Compression gauge		Measures the pressure of the compression. Yanmar code No. TOL-97190080	

4. Disassembly and reassembly

4.2.4 Other material

Items		Usual contents	Features and application
Liquid gasket	Three Bond No.1 TB1101	200g (1 kg also available)	Non-drying liquid gasket ; solvent less type, easy to remove, superior in seawater resistance, applicable to various mating surfaces.
	Three Bond No.2 TB1102	200g (1 kg also available)	Non-drying liquid gasket ; easy to apply, superior in water resistance and oil resistance, especially superior in gasoline resistance.
	Three Bond No.3 TB1103	150g	Drying film, low viscosity and forming of thin film, appropriate for mating surface of precision parts.
	Three Bond No.4 TB1104	200g (1 kg also available)	Semi-drying viscoelastic material, applicable to non-flat surface having many indentations and protrusions, superior in heat resistance, water resistance, and oil resistance.
	Three Bond No.10 TB1211	100g	Solvent-less type silicone-base sealant, applicable to high temperature areas. (-50°C to 250°C)
	Three Bond TB1212	100g	Silicone-base, non-fluid type, thick application possible.
Adhesive	Three Bond TB1401	200g	Prevention of loose bolts, gas leakage, and corrosion. Torque required to loosen bolt : 10 to 20% larger than tightening torque.
	Lock tight SUPER TB1324	50g	Excellent adhesive strength locks bolt semi-permanently.
Seal tape		5m round tape	Sealing material for threaded parts of various pipes. Ambient temperature range : -150°C to 200°C
O-ring kit		Ø 1.92-m dia.:1 Ø 2.42-m dia.:1 Ø 3.12-m dia.:1 Ø 3.52-m dia.:1 Ø 5.72-m dia.:1	O-ring of any size can be prepared, whenever required. (Including adhesive, release agent, cutter, and jig)
EP lubricant (molybdenum disulfate)	Brand name (LOWCOL PASTE)	50g	For assembly of engine cylinders, pistons, metals shafts, etc. Spray type facilitates application work.
	Brand name (PASTE SPRAY)	330g	
	Brand name (MOLYPASTE)	50g	Prevention of seizure of threaded parts at high temperature. Applicable to intake and exhaust valves. (stem, guide, face)

4. Disassembly and reassembly

Items		Usual contents	Features and application
Scale solvent	Scale solvent	1 box (4 kg × 4 removers)	• The scale solvent removes scale in a short time. (1 to 10 hours) • Prepare water (seawater is possible) in an amount that is about 10 times the weight of the solvent. Mix the solvent with water. • Just dipping disassembled part into removes scale. To shorten removal time, stir remover mixture. • If cleaning performance drops, replace remover mixture with new remover mixture. • Neutralize used mixture, and then dispose of it. To judge cleaning performance of mixture, put pH test paper into mixture. If test paper turns red, remover mixture is still effective.
	Neutralizer (caustic soda)	1 box (2 kg × 4 neutralizers)	
	pH test paper		
Antirust			Add antirust to fresh water system. Then operate engine for approximately 5 minutes. Antirust will be effective for 6 months.
Anti freeze			Add anti freeze to fresh water system at the cold area to engine operate.
Cleaning agent			• The cleaning agent removes even carbon adhering to disassembled parts. • If a cleaning machine is used, prepare 4 to 6% mixture of 60° to 80° to ensure more effective cleaning.

[Notice]
It is recommended that the liquid gasket of Three Bond TB1212 should be used for service work.
Before providing service, observe the cautions below:
1) Build up each gasket equally.
2) For a bolt hole, apply liquid gasket to the inside surface of the hole.
3) Conventionally, Three Bond TB1104 (gray) or Three Bond
   TB1102 (yellow) is used for paper packing though single use of one of these bounds is not effective.
4) If conventional packing is used, do not use liquid packing.

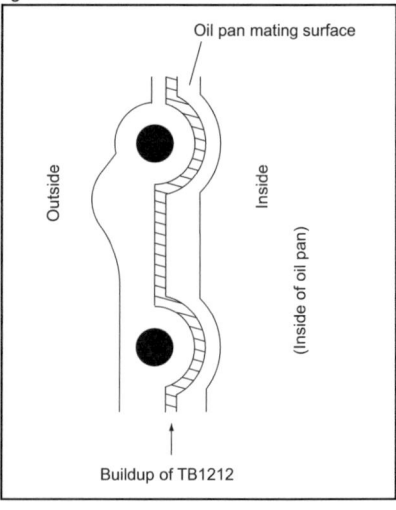

4.3 Disassembly and reassembly

4.3.1 Disassembly

• Preparation on a boat

For engines mounted in an engine room of a ship, remove the piping and wiring connecting them to the ship.
1) Remove the remote control cable (from engine and marine gearbox).
2) Unplug the extension cord for the instrument panel from the engine.
3) Remove the wiring between the starting motor and the battery.
4) Remove the exhaust rubber hose from the mixing elbow.
5) Remove the rubber hose connecting the coolant recovery tank to the filler cap.
6) Remove the seawater inlet hose for the seawater pump (after making sure the seacock is closed).
7) Remove the fuel oil inlet rubber hose from the fuel feed pump.
8) Remove the body fit (reamer) bolts and disassemble the propeller shaft coupling and thrust shaft coupling.
9) If a driven coupling is mounted to the front drive coupling, disassemble.
10) Remove the flexible mount nut, lift the engine, and remove it from the engine base.
    (Leave the flexible mount attached to the engine base.)

• Disassembling an engine in a workshop

(1) Drain cooling water
    Refer to 2.2.5(6).
1) Open a seawater drain cock to drain the seawater.
2) Open a drain cock on the cylinder body to drain the fresh water from the cylinder head and cylinder body.
3) Open a fresh water drain cock on the lower part of the coolant tank to drain the fresh water.
4) Open a fresh water drain cock on the lower part of the fresh water pump to drain the fresh water.

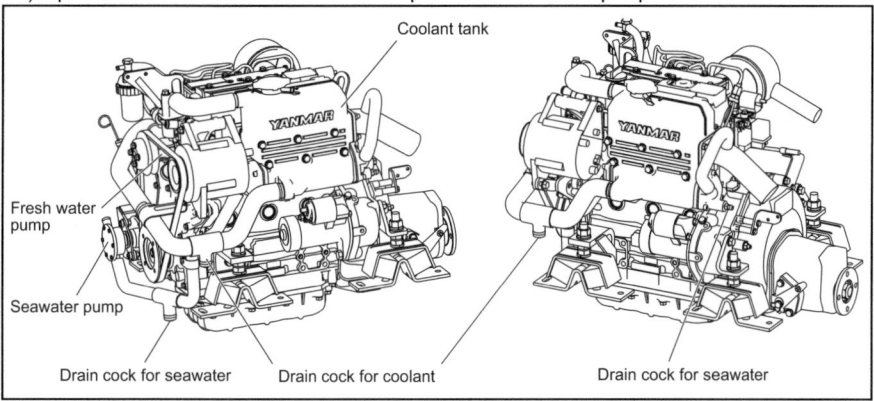

(2) Drain lube oil
1) Remove the pipe coupling bolt, which holds the lube oil dipstick guide, and drain the lube oil from the engine.
2) Remove the drain plug on the lower part of the crank case control side, and drain the lube oil from the marine gearbox.

4. Disassembly and reassembly

[NOTE:]
If a lube oil supply/discharge pump is used for the engine, the intake hose is placed in the dipstick guide, and for the clutch side (gearbox) it is placed in the oil hole on top of the case. (Refer to 2.2.2(2) & (3))

(3) Removing (electrical) wiring
Remove the wiring from the engine. (Refer to 2.2.4(9))

(4) Removing the fuel filter & fuel pipe
1) Remove the fuel pipes (fuel filter-fuel feed pump, fuel filter-fuel injection pump and fuel nozzle-fuel pump)
2) Remove the fuel filter.

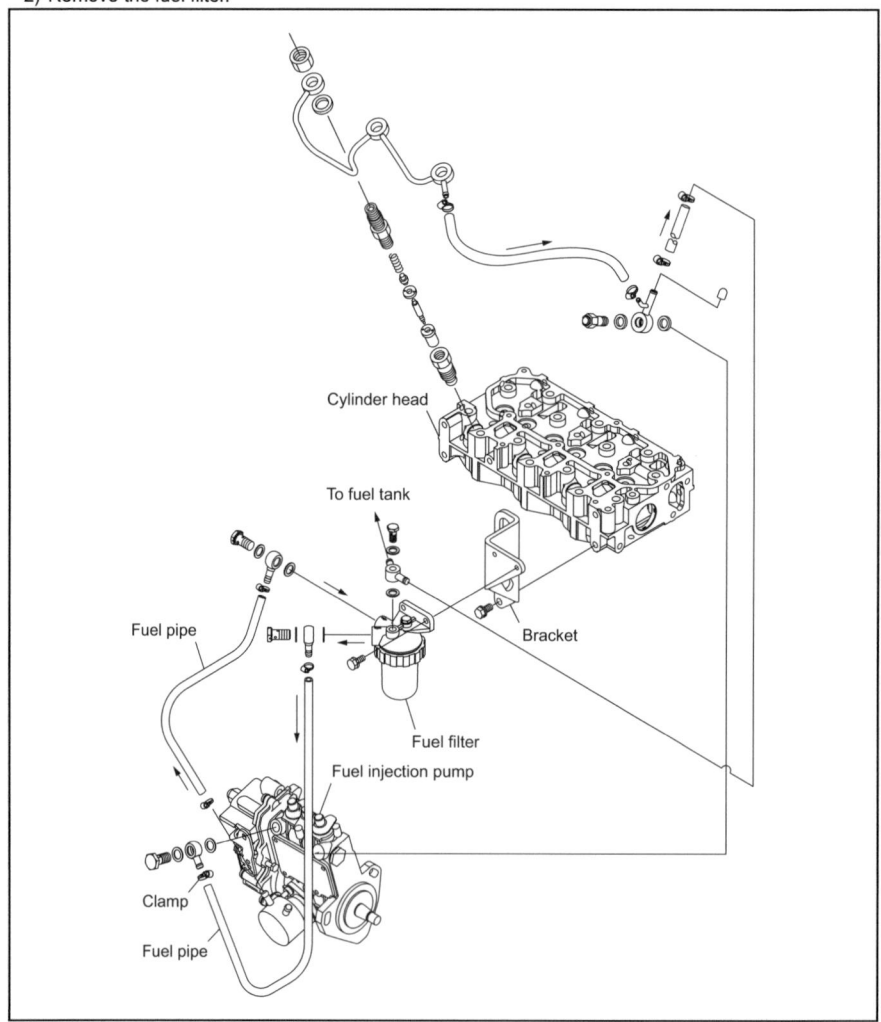

4. Disassembly and reassembly

(5) Removing the intake silencer
1) Remove the air intake hose attached to the intake manifold.
2) Remove the intake silencer from cylinder head.

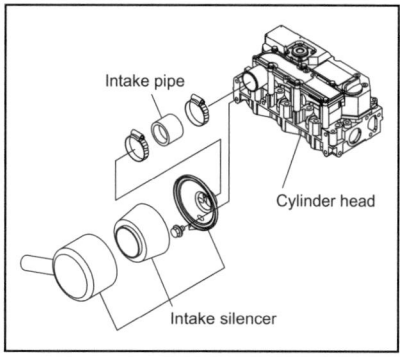

(6) Removing the mixing elbow
1) Remove the seawater rubber hose connecting to a heat exchanger- a mixing elbow.
2) Remove the mixing elbow from the exhaust manifold.

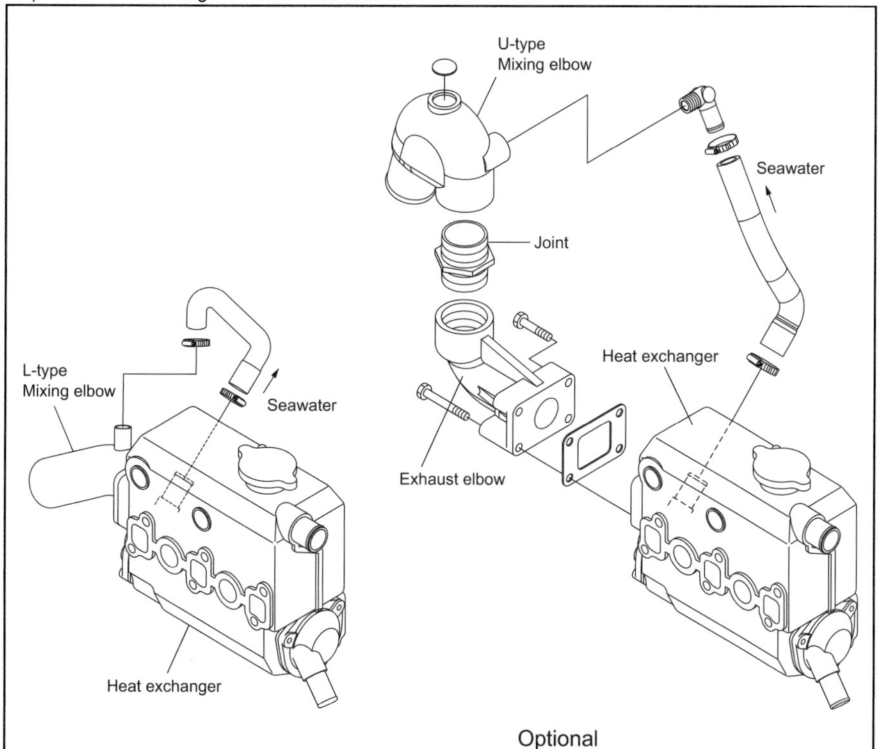

(L-type mixing elbow)

Optional
(U-type mixing elbow)

4. Disassembly and reassembly

(7) Removing the alternator

1) Loosen the set bolt (belt adjuster bolt) and move the alternator downwards. Loosen the V-pulley set bolts for the cooling water pump and remove the V-belt.
2) Remove the belt adjuster from the gear case and remove the alternator from the cylinder head (with bracket).

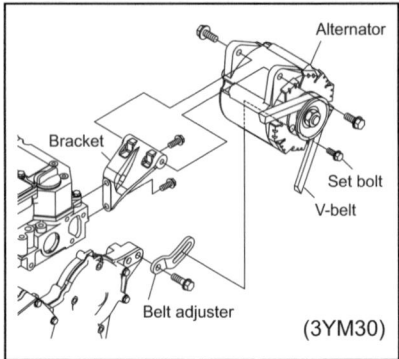

(3YM30)

(8) Removing the cooling water pipes (seawater/ fresh water)

1) Remove the seawater pipe (seawater pump-heat exchanger).
2) Remove the fresh water pipe (fresh water pump - heat exchanger, coolant tank - fresh water pump).

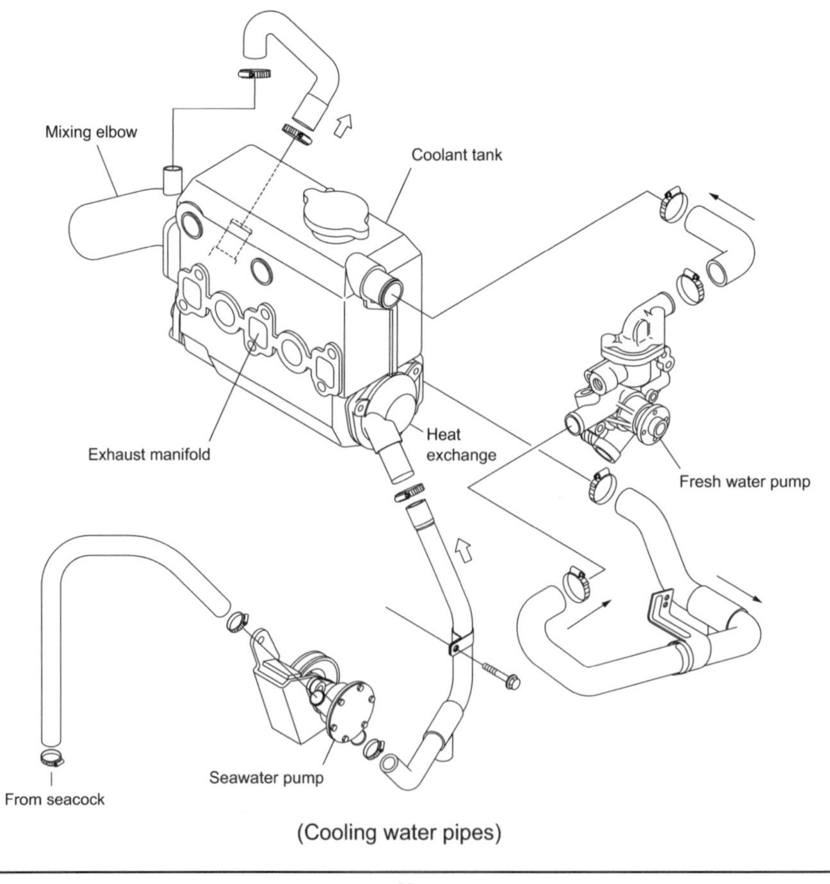

(Cooling water pipes)

4. Disassembly and reassembly

(9) Removing the heat exchanger (exhaust manifold, coolant tank unit)
Remove the heat exchanger and gasket packing from cylinder head.

(10) Removing the seawater pump
Loosen the nut on the spacer bolt and remove the seawater pump.

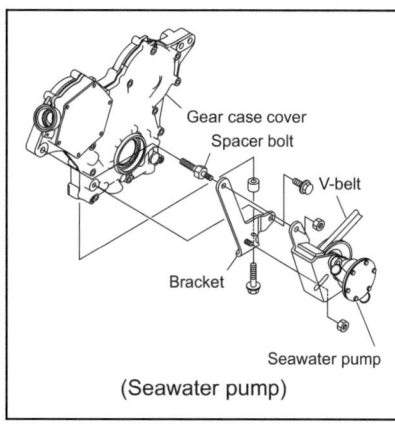
(Seawater pump)

(11) Removing the starting motor
Remove the starting motor from the flywheel housing.

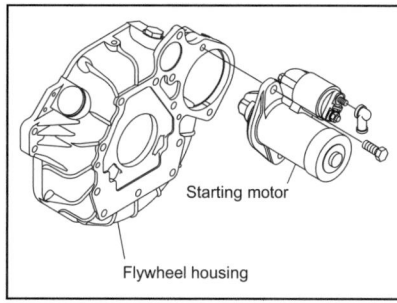

(12) Removing the lube oil filter
Remove the lube oil filter from the cylinder block.

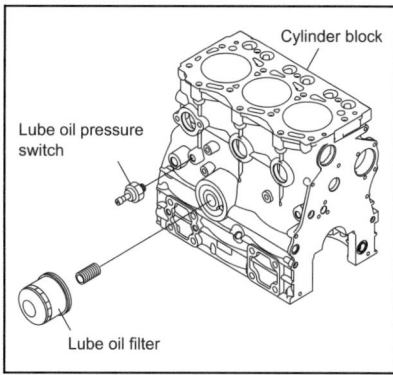

4. Disassembly and reassembly

(13) Removing the fuel injection pipe
1) Remove the fuel injection pipe retainer.
2) Loosen the cap nuts on both ends of the fuel injection pipe and remove the fuel injection pipe.

Note:
When loosening the cap nut, loosen the nut with holding the return pipe by hand so that the fuel return pipe may not break.

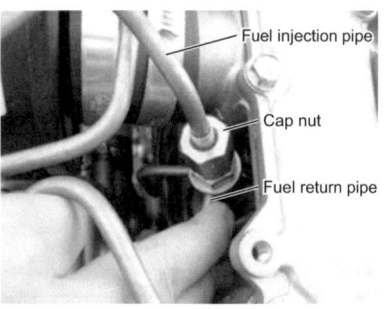
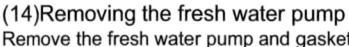
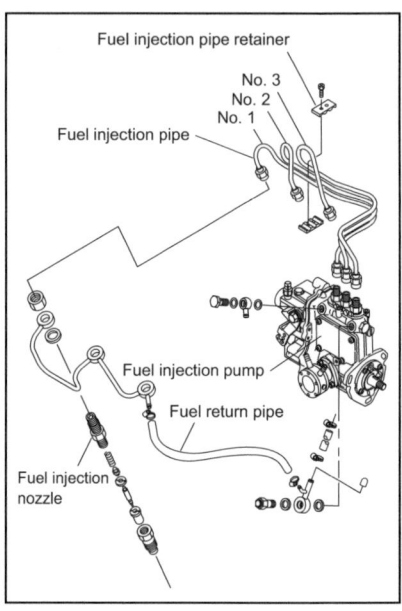

3) Remove the fuel return pipe (fuel nozzle-fuel pump)

(14) Removing the fresh water pump
Remove the fresh water pump and gasket.

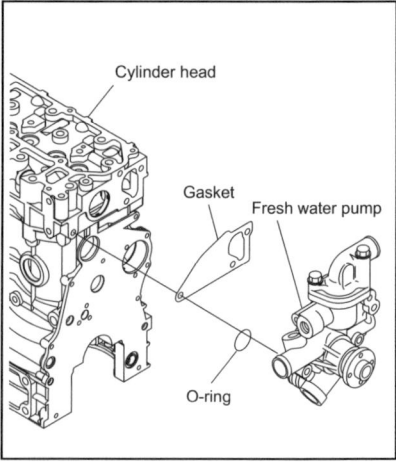

4. Disassembly and reassembly

(15) Removing the fuel injection nozzles
Loosen the fuel nozzles and remove them from the cylinder head.

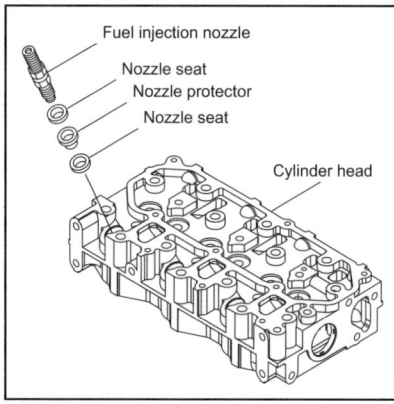

(16) Remove the fuel injection pump
1) Loosen the drive gear nut for fuel pump drive gear, and pull out the fuel pump drive gear /flange assembly with an extraction tool.

Note:
Don't disassemble the pump flange and the pump drive gear.

2) Loosen the three (3) nuts for the fuel pump and remove the fuel injection pump and O-ring from the gear case.

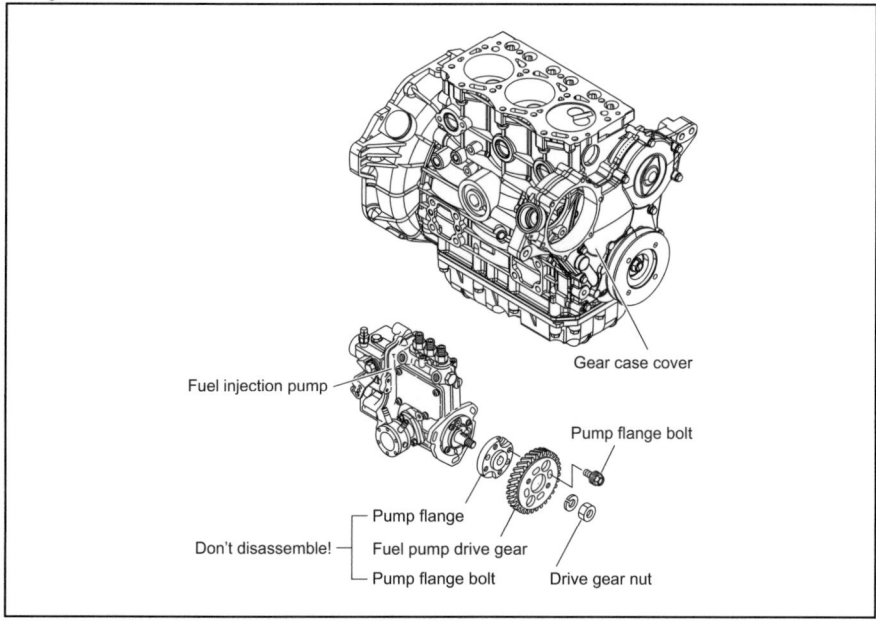

4. Disassembly and reassembly

(17) Removing the rocker arm shaft assembly

1) Loosen the hexagon head bolts and remove the rocker arm cover.

Note:
Don't loosen the cross-recessed-head bolts on the rocker arm cover, when removing the rocker arm cover.

2) Remove the bolts(s) for the rocker arm shaft support, and remove the entire rocker arm shaft assembly.
3) Pull out the push rods.

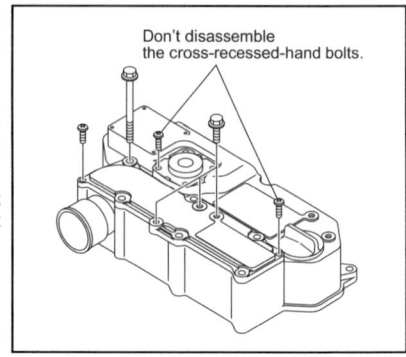

Don't disassemble the cross-recessed-hand bolts.

(18) Removing the cylinder head

1) Loosen the cylinder head bolts with a torque wrench, and remove the cylinder head.
2) Remove the cylinder head gasket.

(19) Removing the marine gearbox

Loosen the bolts for the clutch case flange, and remove the gearbox assembly.

(20) Removing the flywheel

Loosen the flywheel bolts and remove the flywheel.

Note:
Be careful not to scratch the ring gear.

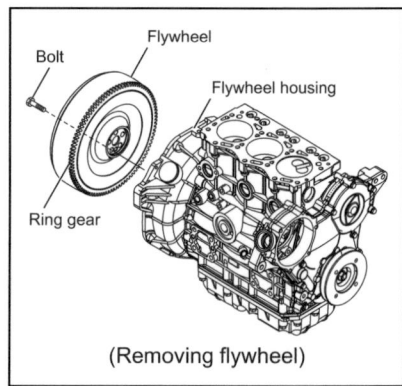

(Removing flywheel)

4. Disassembly and reassembly

(21) Turning the engine over
1) Place a wood block of appropriate size on the floor, and stand up the engine on the flywheel housing.
2) Remove the engine mounting feet.

(22) Removing the crankshaft V-pulley
Loosen the bolt tightening the crankshaft V-pulley and remove the crankshaft V-pulley with a pulley puller.

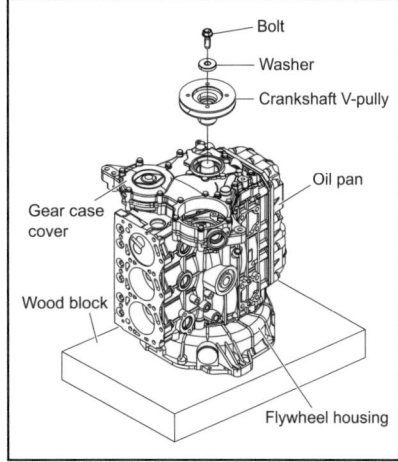

(23) Removing the oil pan
Remove the oil pan and spacer.

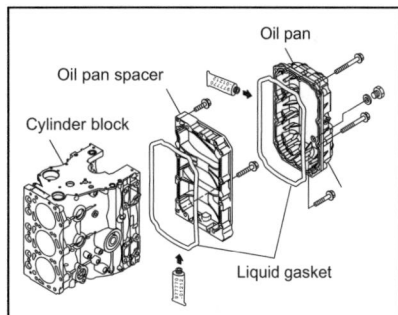

(24) Removing the lube oil inlet pipe
Remove the lube oil inlet pipe and O-ring.

(25) Removing the gear case
Loosen the gear case cover bolts, and remove the gear case cover from the gear case.

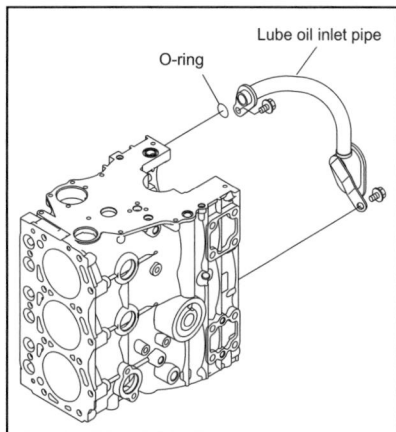

(26) Removing the lube oil pump
Remove the lube oil pump from the gear case cover.

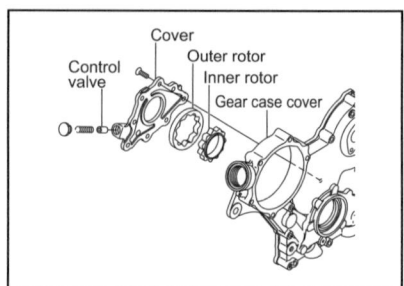

(27) Removing the idle gear
Loosen the three bolts holding the idle gear and pull out the idle gear and shaft.

(28) Removing the camshaft
1) Push up tappet by turning a camshaft to remove it from the cylinder block easily.
2) Loosen the thrust metal bolts through the holes of the camshaft gear, and remove.
3) Pull out the camshaft gear and camshaft assembly from the cylinder block.

NOTE:
The camshaft gear and camshaft are shrunk fit. They must be heated to 180-200 °C to disassemble.

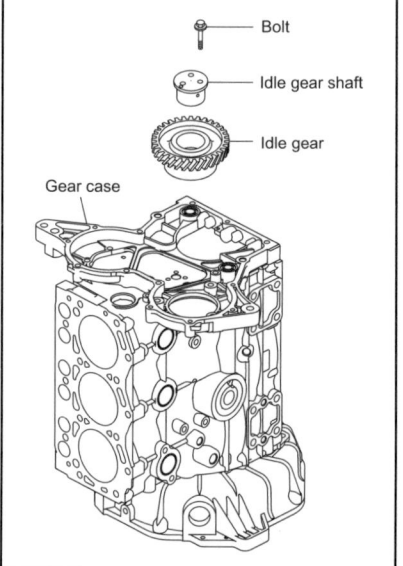

(29) Removing the pistons and connection rods
1) Loosen the rod bolts and remove the large end cap.
2) Push the connecting rod and pull out the piston & connecting rod assembly.

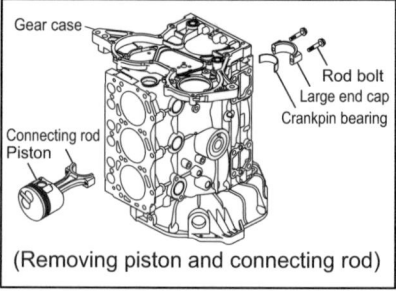

(Removing piston and connecting rod)

4. Disassembly and reassembly

(30) Removing the gear case
1) Remove the gear case from the cylinder block.
2) Remove the O-rings from the lube oil passage.

NOTE
1) When mounting the gear case, match up the two knock pins for cylinder block.
2) Be sure to coat the O-rings for the cylinder block lube oil passage with grease when assembling, so that it does not get out of place.

(31) Loosening the main bearing bolts.
Loosen the main bearing bolts. Don't remove them.

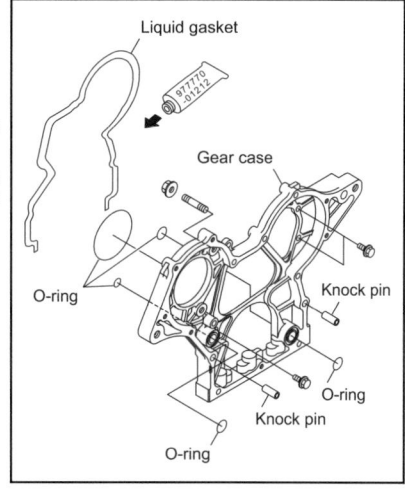

(32) Turning the engine over
Turn the engine over, with the cylinder head mounting surface facing down.

NOTE:
Make sure that the cylinder head positioning pins on the cylinder block do not come in contact with the wood block.

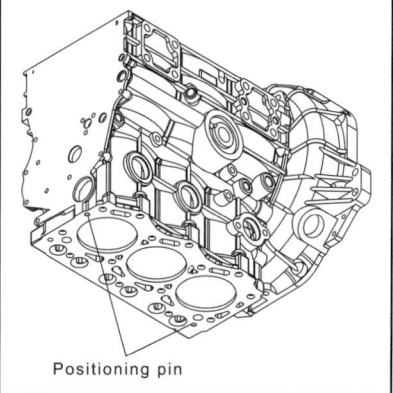

(33) Removing the flywheel housing
Remove the flywheel housing with the oil seal from the cylinder block.

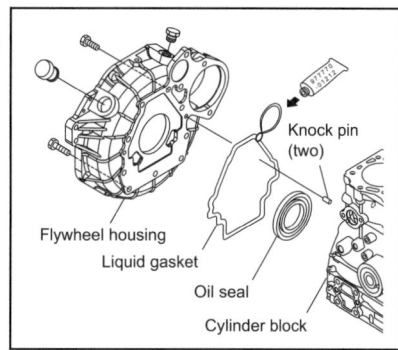

4. Disassembly and reassembly

(34) Removing the main bearing
1) Remove the main bearing bolts.
2) Remove the main bearing cap and lower main bearing metal.

NOTE:
The thrust metal (lower) is mounted to the base main bearing cap.

(35) Removing the crankshaft
1) Remove the crankshaft

NOTE:
1) The thrust metal (upper) is mounted to the standard main bearing of the cylinder block.
2) Remove the main bearing metal (upper) from the cylinder block.

(36) Removing the tappets
Remove the tappets from the tappet holes in the cylinder black.

NOTE:
1) Be careful to keep all disassembled parts in order.
2) Prepare clear and adequate area for parts and a container(s).
3) Prepare a cleaner before disassembling.

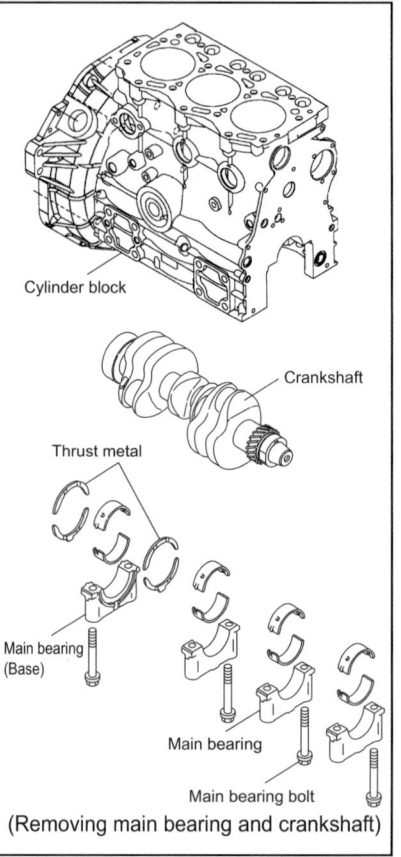

(Removing main bearing and crankshaft)

4.3.2 Reassembly

(1) Clean all parts
Clean all parts using by the cloth and diesel oil (or cleaning agent) before reassembly.

NOTE:
1) If the dust remain with the parts, engine may cause the seizing or damage.
2) The cleaning agent removes even carbon adhering to disassembled parts.

(2) Putting the cylinder block upside down
Place a wood block on the floor and put the cylinder block upside down (with the cylinder head mounting surface facing down).

(3) Inserting the tappets
Coat the inside of the cylinder block tappet holes and the outside circumference of the tappets with engine oil, and insert the tappets in the cylinder block.

NOTE:
Separate the tappets to make sure that they are reassembled in the same cylinder No. and intake/exhaust side as they came from.

(4) Mounting the crankshaft
1) The crankshaft and the crankshaft gear are shrink fitted. If the crankshaft and the crankshaft gear have been disassembled, they have to be shrink fitted [heat the crank shaft gear to 180- 200 deg. C in the hot oil and press fit to the crankshaft].
2) Coat the crank journal part of the cylinder block and the upper main bearing metal with oil and fit the upper main bearing metal onto the cylinder block.

NOTE:
1) Be sure not to confuse the upper and lower main bearing metals. The upper metal has an oil groove.
2) When mounting the thrust metal, fit it so that the surface with the oil groove slit faces outwards, crankshaft side.

3) Coat the crankpins and the crank journals of the crankshaft with engine oil and place it on the upper main bearing metals.

NOTE:
1) Position so that the crankshaft gear is on the gear case side.
2) Be careful not to let the thrust metal drop.

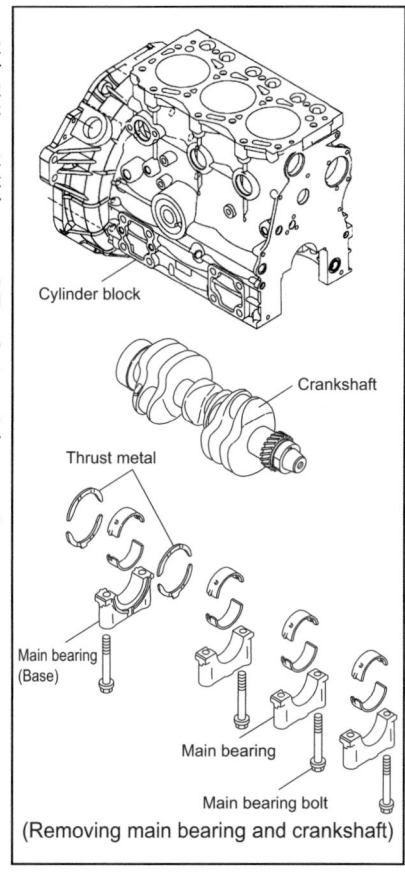

(Removing main bearing and crankshaft)

4. Disassembly and reassembly

(5) Mounting the main bearing cap

Coat the lower main bearing metal with engine oil, and mount it to the main bearing cap.

NOTE:
1) The lower main bearing metal does not have an oil hole.
2) The base bearing thrust metal is fitted with the oil groove facing outwards.

1) Coat the flange and the thread of the main bearing bolts with engine oil, put them on the crankshaft journal, and tighten the main bearing bolts to the specified torque.

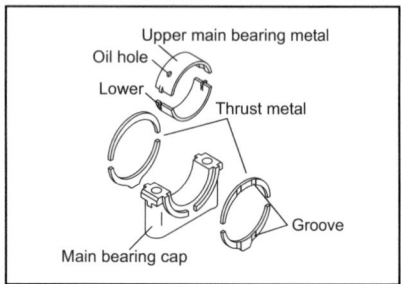

Tightening torque	N·m (kgf·m)
Main bearing bolt	75.5-81.5 (77-8.3)

NOTE:
1) The main bearing cap should be mounted with the arrow on the cap pointing towards the flywheel.
2) Make sure to have the correct cylinder alignment number.

2) Measure the crankshaft side clearance.

		mm
Crankshaft side clearance	Standard	Limit
	0.133-0.233	0.28

3) Make sure that the crankshaft rotates smoothly and easily.

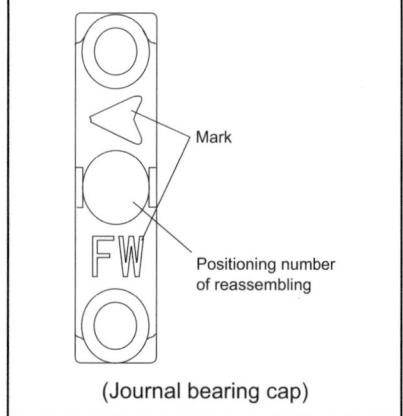

(Journal bearing cap)

(6) Mounting the flywheel housing

1) Replace the used oil seal with new one. Press fit the oil seal in the flywheel housing, and coat the lip of the oil seal with engine oil.
2) Apply the liquid gasket on the mounting surface of the flywheel housing and mount the flywheel housing to the cylinder block, while matching up with the knock pins.

NOTE:
Be careful that the liquid gasket does not protrude onto the oil pan mounting surface.

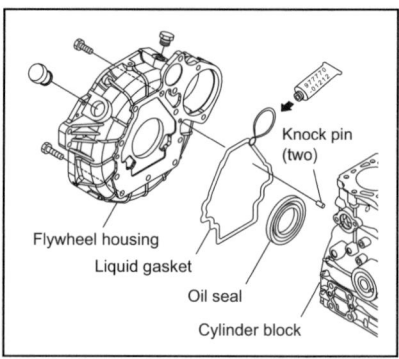

4. Disassembly and reassembly

(7) Turning the engine over
Stand up the engine on the flywheel housing.

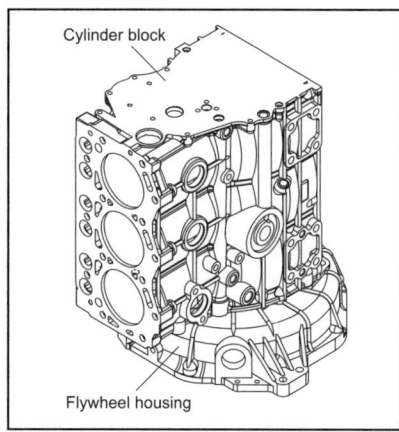

(8) Mounting the gear case
Apply the liquid gasket to the gear case and mount the gear case and lube oil line O-ring onto the cylinder block.

NOTE:
1) When mounting the gear case, match up the two knock pins of the cylinder block.
2) Be sure to coat the O-ring for the cylinder block lube oil line with grease when assembling, so that it does not get out of place.

(9) Mounting the piston and connecting rod
1) Reassemble the piston and connecting rod.

   NOTE:
   1) When reassembling the piston and connecting rod, make sure that the parts are assembled with the correct orientation.
   2) Install each piston ring on the piston, with the punched manufacturer's mark facing upward.
   3) The piston ring joints shall be staggered at by 120° intervals. Do not position the top ring joint vertical to the piston pin. The coil expander joint shall be opposite to the oil ring joint.

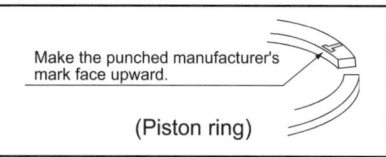

(Piston ring)

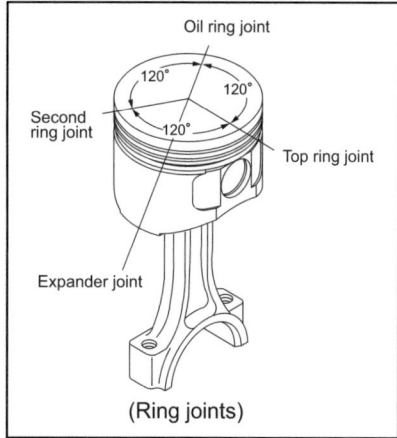

(Ring joints)

4. Disassembly and reassembly

2) Coat the outside of the piston and the inside of the connecting rod crank pin metal with engine oil and insert the piston with the piston insertion tool.

NOTE:
1) Insert the piston so that the match mark on the large end of the connecting rod faces the fuel nozzle, and the manufacture's embossed mark on the stem faces toward the flywheel.
2) After inserting the piston, make sure the ID mark on the piston top is located at the camshaft side, looking from the top of the piston.
3) Align the large end match mark, mount the cap, and tighten the connecting rod bolts.

Tightening torque	N·m (kgf·m)
Connecting rod bolt	22.6-27.5(2.3-2.8)

NOTE:
If a torque wrench is not available, match up with the mark made before disassembly.

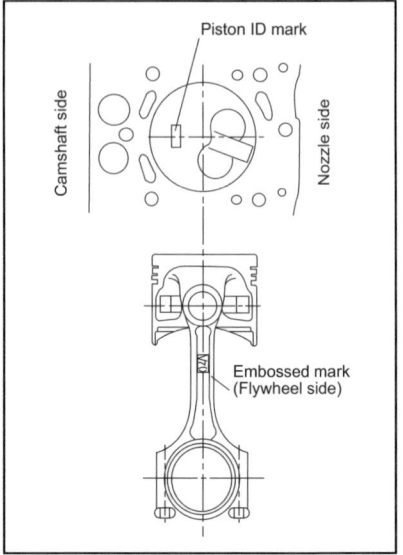

(10) Mounting the camshaft

1) If the camshaft and the camshaft gear have been disassembled, shrink fit the camshaft and the camshaft gear [heat the camshaft gear to 180-200 deg. C in the hot oil and press fit].

NOTE:
When mounting the camshaft and the camshaft gear, be sure not to forget assembly of the thrust metal. Also make sure they are assembled with the correct orientation.

2) Coat the cylinder block camshaft bearings and camshaft with engine oil, insert the camshaft in the cylinder block, and tighten the thrust metal bolts.

3) Measure the camshaft side gap.

mm

Camshaft side gap	Standard	Limit
	0.05-0.15	0.25

4) Make sure that the camshaft rotates smoothly.

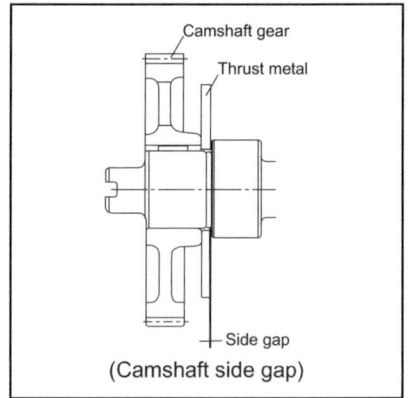

(Camshaft side gap)

4. Disassembly and reassembly

(11) Mounting the idling gear

1) Mount the idling gear so that the mark of the idle gear shaft upward.

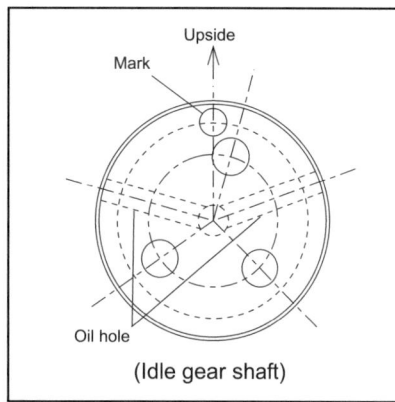
(Idle gear shaft)

2) Align the "A" and "B" match marks of the idle gear with the match marks of the crankshaft gear and the camshaft gear.
3) Measure the idle gear, camshaft gear and crankshaft gear backlash.

mm

Backlash	Standard	Limit
	0.07-0.15	0.17

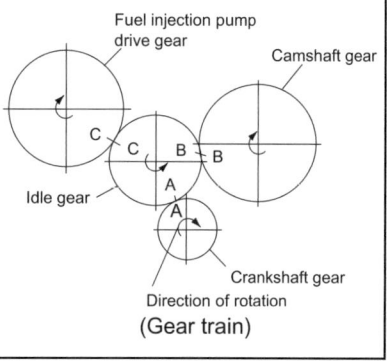
(Gear train)

(12) Mounting the lube oil pump

Tighten the lube oil pump with bolts to the gear case cover.

Lube oil applied   N•m (kgf•m)

Tightening torque	5.9-7.9 (0.6-0.8)

NOTE:
1) Before installing the outer /inner rotors, coat them with lube oil (10W30 lube oil).
2) Assemble the rotor so that the mark of the rotor may come to the cover side.
3) Confirm that the rotor rotates smoothly.

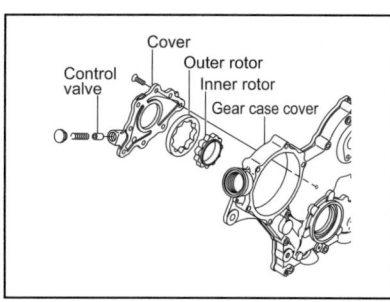

4. Disassembly and reassembly

(13) Mounting the oil seal and gear case cover

1) Replace the used oil seal with a new one when the gear case cover is disassembled.
2) Insert a new oil seal by using the oil seal insertion tool on the position of the gear case cover end face (61.5 0/-1 mm distance from the end of the cylinder block). (Refer to the right figure.)
3) Apply lithium grease to the oil seal lips.
4) When wear is found on the oil seal contact part of a crankshaft pulley, replace the pulley with a new one. Carefully install the pulley so as not to damage the oil seal.
5) Apply the liquid gasket to the gear case cover. Position the two knock pins and tighten the bolts of the gear case cover.

NOTE:
Trim the liquid gasket if it protrudes onto the oil pan mounting surface.

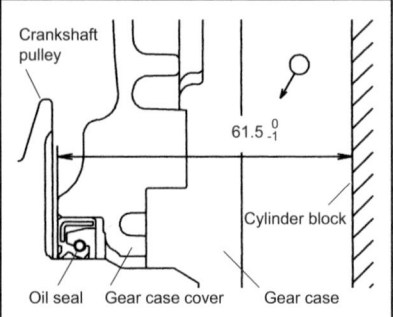

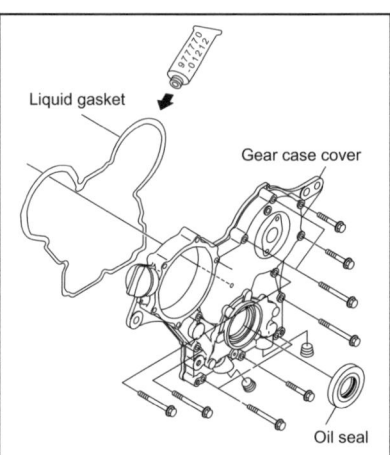

(14) Mounting the lube oil inlet pipe

Mount the lube oil inlet pipe on the bottom of the cylinder block, using new gasket.

Tightening torque	N·m (kgf·m)
Lube oil inlet pipe	26 (2.6)

94R1

4. Disassembly and reassembly

(15) Mounting the oil pan spacer and the oil pan

1) Apply the liquid gasket to the surfaces of the gear case cover, gear case and flywheel housing that contact with the cylinder block.
2) Apply the liquid gasket to the spacer. Mount the spacer to the cylinder block and tighten the bolts.
3) Apply the liquid gasket to the oil pan. Mount the oil pan to the spacer and tighten the bolts.
4) Mount the dipstick and dipstick guide.

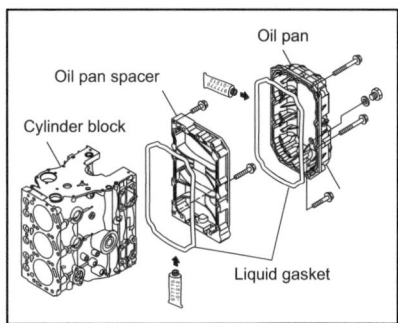

(16) Mounting the crankshaft pulley

1) Coat the oil seal with oil.
2) Make sure to wipe off oil on the taper surfaces of the crankshaft and the pulley.
3) Mount the pulley insertion tool to the crankshaft and install the pulley on the crankshaft so that the oil seal may not be damaged.
4) Tighten the pulley to the specified torque.

Tightening torque	N·m (kgf·m)
V-pulley bolt (Material : casting iron)	83.3-93.3 (8.5-9.5)

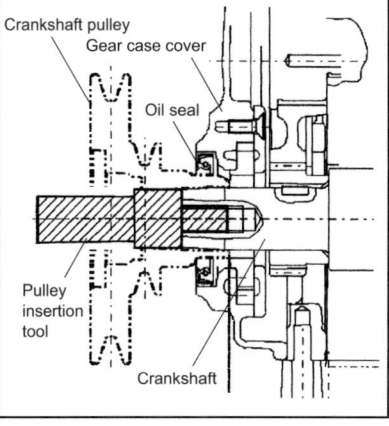

(17) Mounting the engine mounting feet and turning the engine upright.

1) Mount the engine mounting feet.
2) Turn the engine upright (Oil pan is the bottom side).

(18) Mounting the flywheel

1) Coat the flywheel bolts threads with engine oil.
2) Align the positioning pins, and tighten the flywheel bolts to the specified torque.

Tightening torque	N·m (kgf·m)
Flywheel bolt	80.4-86.4 (8.2-8.8)

4. Disassembly and reassembly

(19) Mounting the marine gearbox

1) Mount the damper disk to the flywheel.
   Align the damper disk with the input shaft spline and insert.
2) Mount the marine gearbox to the flywheel housing.

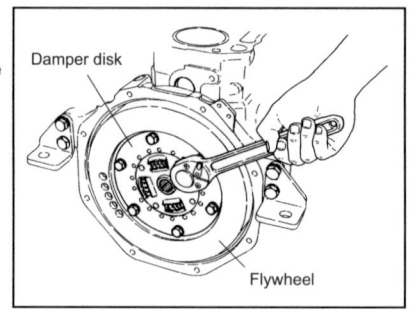

(20) Mounting the cylinder head

1) Put the cylinder head gasket on the cylinder block, aligning it with the cylinder block positioning pins.
2) Lift the cylinder head horizontally and mount it aligning with the cylinder head gasket.
3) Coat the flange part and thread of the cylinder head bolt with engine oil, and lightly tighten the bolts in the specified order first. Then tighten completely, in the same order.

Tightening torque   N·m (kgf·m)

	1st step	Final
Cylinder head bolt	24-32 (2.5-3.3)	53.9-57.9 (5.5-5.9)

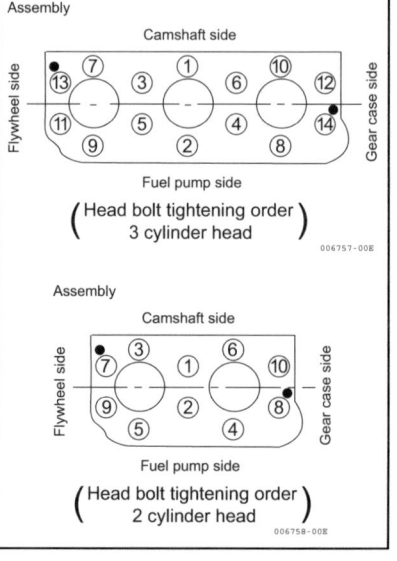

(21) Mounting the rocker arm shaft assembly and pushrods

1) Fit the pushrod to the tappet.
2) Coat the top of the pushrod and the adjusting screw of the rocker arm with engine oil. (Apply lube oil to the screw and lock nut.)
   Mount the rocker arm shaft assembly to the cylinder head and tighten the bolts.
3) Adjust valve clearance

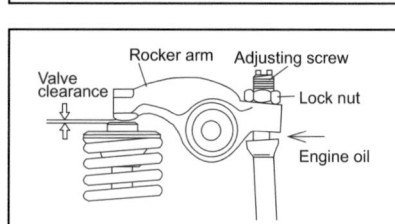

mm

Intake/exhaust valve clearance	0.15-0.25

4) Coat the rocker arm and valve spring with engine oil and mount the rocker arm cover.

4. Disassembly and reassembly

(22) Mounting the fuel injection pump

1) Lightly fit the fuel injection pump on the gear case. (After adjusting injection timing, tighten the fuel injection pump. Before mounting the heat exchanger to the cylinder head, adjust the injection timing.).

   NOTE:
   Be careful not to scratch the O-ring between the fuel injection pump and gear case.

2) Fit the fuel injection pump drive gear to the fuel pump camshaft.

3) Align the "C" match marks on the fuel injection pump drive gear and idle gear.

4) Tighten the pump drive gear nut to the specified torque. (Do not apply lube oil to the nut.)

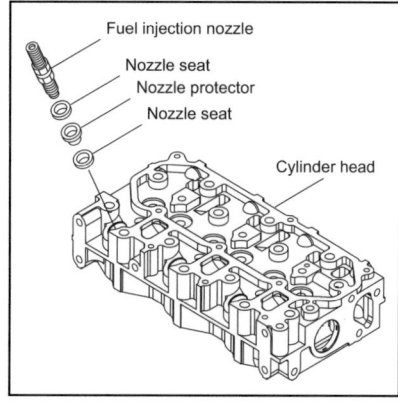

(Gear train)

Tightening torque	N•m (kgf•m)
Pump drive gear nut	58.9-68.7 (6.0-7.0)

5) Measure the backlash of the fuel injection pump drive gear.

Backlash	mm
Fuel injection pump drive gear	0.06-0.12

(23) Mounting the fuel injection nozzle

1) Replace the used fuel nozzle protector and fuel nozzle seat with new ones. Put the seat in the cylinder head and the protector to the nozzle tip. Mount the fuel injection nozzle to the cylinder head.

2) Tighten the fuel nozzle retainer bolt to the specified torque. (Do not apply lube oil to the bolt.)

Tightening torque	N•m (kgf•m)
Fuel nozzle retainer bolt	49.0-53.0 (5.0-5.4)

4. Disassembly and reassembly

(24) Mounting the fresh water pump

1) Thoroughly coat both sides of the gasket with adhesive.
2) Renew the O-ring for the connecting part of the pump, which is inserted in the cylinder block, and tighten the fresh water pump.

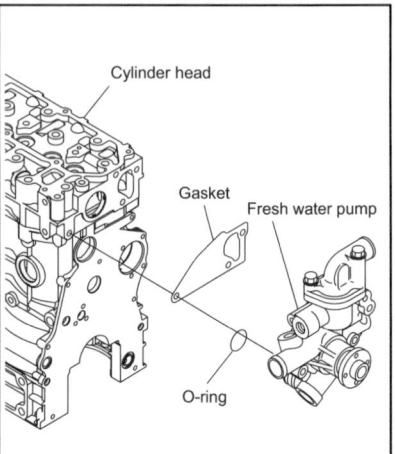

(25) Mounting the fuel injection pipe and fuel return pipe

1) Mount the fuel injection pipe and then assemble the fuel injection pipe retainer to prevent the pipe vibration.

   NOTE:
   Lightly tighten the pipe joint nuts on both ends of the fuel injection pipe. Completely tighten after adjusting the injection timing.

Standard tightening torque	N·m (kgf·m)
fuel injection pipe joint nut	29.4-34.4 (3.0-3.5)

2) Mount the fuel return pipe with the clamp (fuel injection nozzle-fuel injection pump).

   NOTE:
   When tightening the fuel return pipe to fuel injection nozzle, tighten the nut with holding the return pipe by hand so that the pipe may not break. (Refer to 4.3.1(13).)

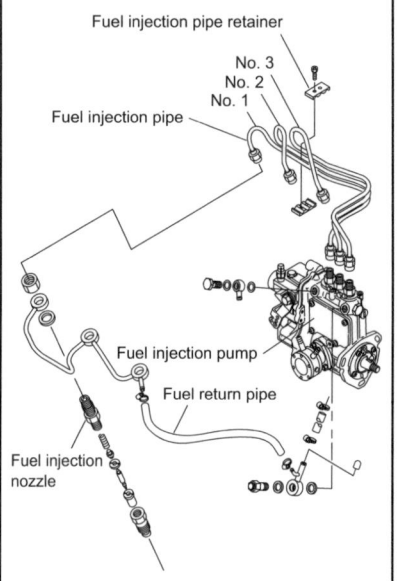

4. Disassembly and reassembly

(26) Mounting the lube oil filter.
Mount the lube oil filter with the tool of the filter case remover.

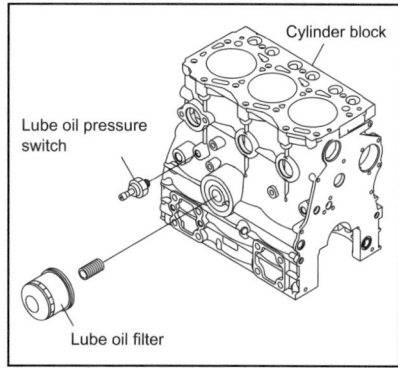

(27) Mounting the seawater pump
1) Tighten the spacer bolt to the gear case cover.
2) Mount the bracket and the seawater pump assembly to the gear case cover.

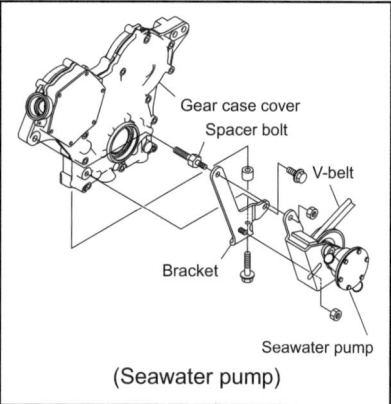

(Seawater pump)

(28) Mounting the heal exchanger (exhaust manifold, fresh water tank unit).
Mount the gasket and heal exchanger (exhaust manifold).

Note:
After adjusting injection timing and tightening the fuel injection pump, mount the heat exchanger. Because it is harder to tighten the fuel pump nuts after installing the heat exchanger.

4. Disassembly and reassembly

(29) Mounting the mixing elbow
1) Mount the mixing elbow on the exhaust manifold outlet.
2) Mount the cooing seawater pipe (rubber hose) with the hose clips (heat exchanger-mixing elbow).

(30) Mounting the cooling water pipes (seawater / fresh water)
1) Mount the seawater pipes with the hose clips (seawater pump-heat exchanger).
2) Mount the fresh water pipes with the hose clips (exhaust manifold - fresh water pump, fresh water pump-heat exchanger).

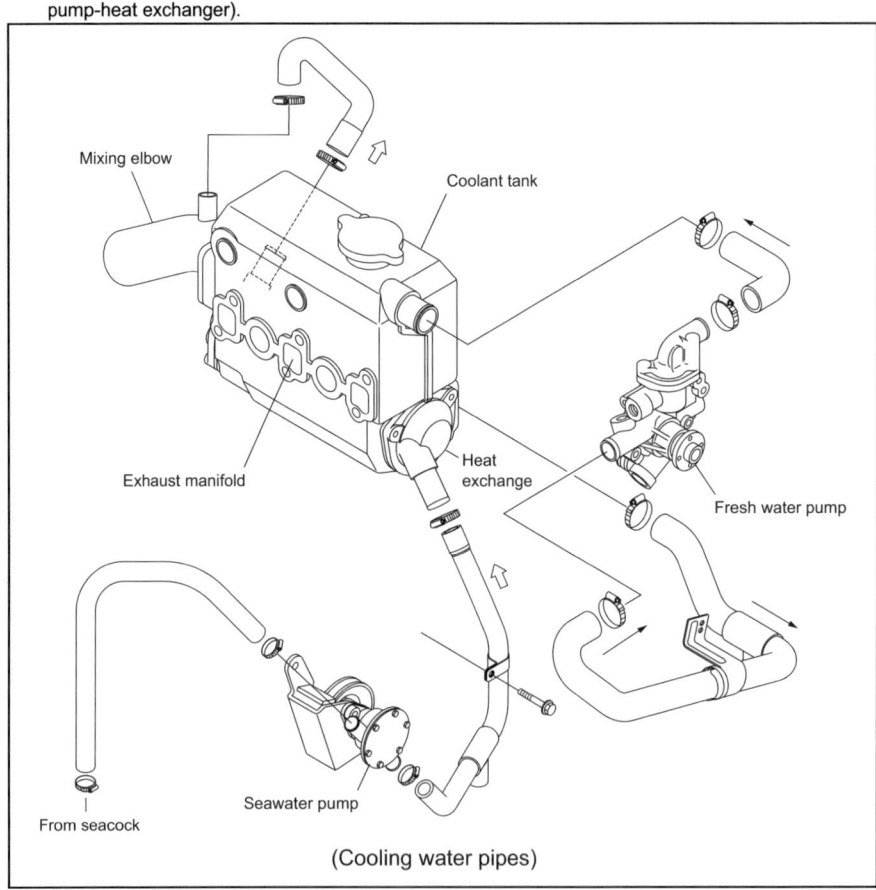

(Cooling water pipes)

4. Disassembly and reassembly

(31) Mounting the alternator
1) Mount the bracket on the cylinder head and the adjuster on the gear case cover, and then the alternator.
2) Adjust V-belt tension with the adjuster, and tighten the set bolt.

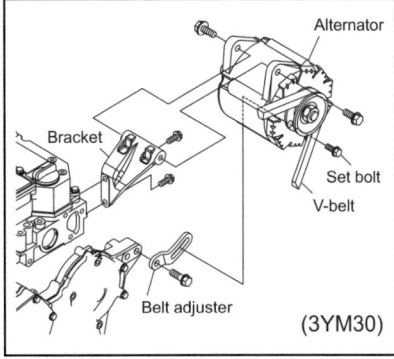

(3YM30)

(32) Mounting the starting motor
Fit the starting motor on the flywheel housing.

(33) Mounting the intake silencer
Mount the intake pipe on the intake manifold inlet coupling and tighten the intake silencer to the cylinder head.

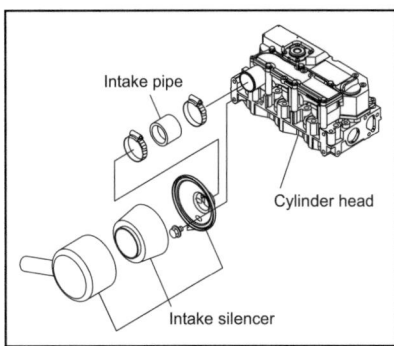

4. Disassembly and reassembly

(34) Mounting the fuel filter and fuel pipe
1) Mount the fuel filter on the bracket, which is tightened to the cylinder head.
2) Mount the fuel pipe (fuel feed pump-fuel filter, fuel filter-fuel injection pump).

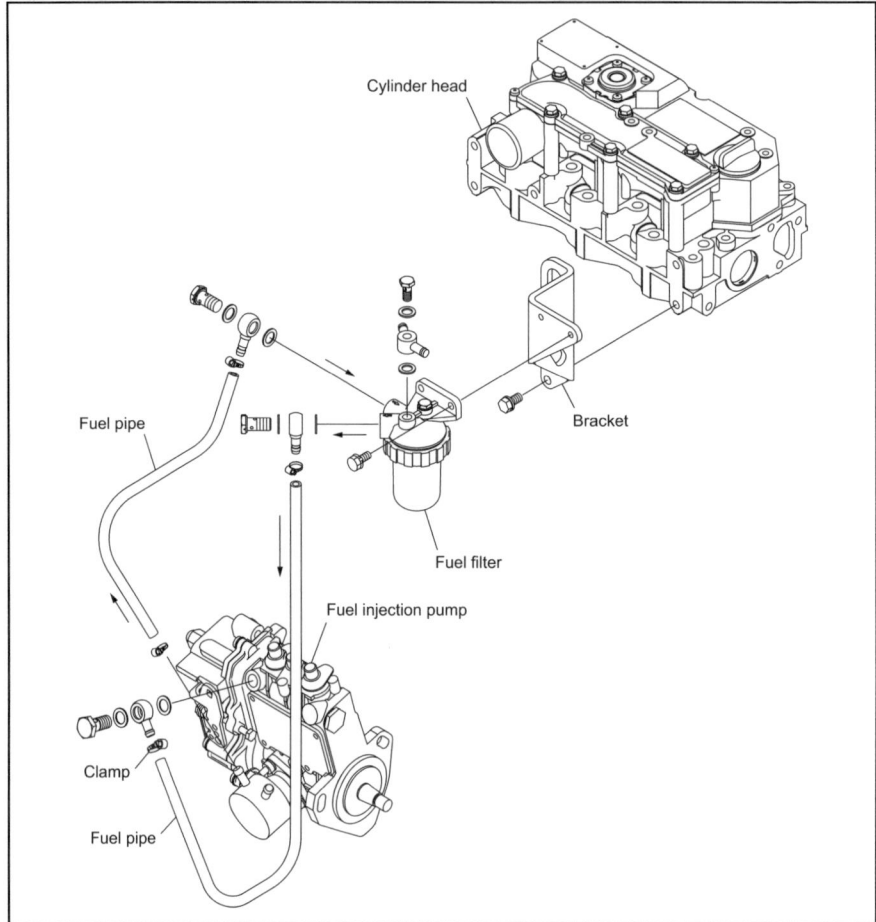

(35) Electrical Wiring
Connect the wiring to the proper terminals, observing the color coding to make sure the connections are correct.

(36) Installation in a boat and completion of the piping and wiring
Mount the engine on the engine bed in the engine room of a boat after all engine assembly has been completed. Connect the cooling water pipes, fuel pipes, other pipes on the boat and the exhaust hoses. Connect the battery, instrument panel, remote control cable and other wiring.

(37) Filling with lube oil
Fill the engine with lube oil from the filler port on top of the gear case or the rocker arm cover. Fill the marine gearbox from the filler port on top of the clutch case.

4. Disassembly and reassembly

(38) Filling with cooling water
1) Open the coolant (fresh water) tank cap and fill with water.

Model	Engine capacity L (quart)
3YM30(C)	4.9 (5.2)
3YM20(C)	4.1 (4.3)
2YM15(C)	3.0 (3.2)

2) Fill with water until the level in the coolant recovery tank is between the full and low marks.

Coolant recovery tank capacity (full)	0.8 L (0.8 quart)

(39) Test running
Refer to "Adjusting operation" of 2.6 in chapter 2.

5. Inspection and servicing of basic engine parts

5.1 Cylinder block

The cylinder block is a thin-skinned, (low-weight), short skirt type with rationally placed ribs. The side walls are save shaped to maximize rigidity for strength and low noise.

5.1.1 Inspection of parts

Make a visual inspection to check for cracks on engines that have frozen up, overturned or otherwise been subjected to undue stress. Perform a color check on any portions that appear to be cracked, and replace the cylinder block if the crack is not repairable.

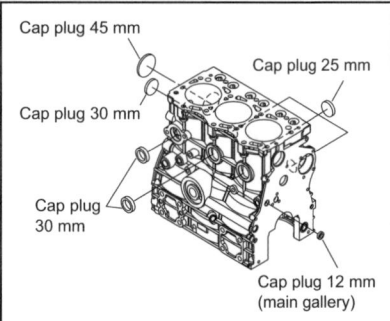

5.1.2 Cleaning of oil holes

Clean all oil holes, making sure that none are clogged up and the cap plugs do not come off.

Color check kit

	Quantity
Penetrant	1
Developer	2
Cleaner	3

5.1.3 Color check procedure

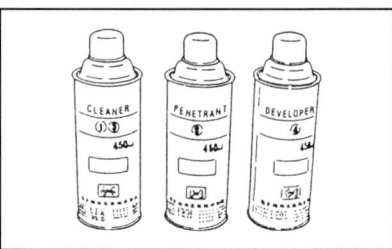

(1) Clean the area to be inspected.

(2) Color check kit
The color check test kit consists of an aerosol cleaner, penetrant and developer.

(3) Clean the area to be inspected with the cleaner.
Either spray the cleaner on directly and wipe, or wipe the area with a cloth moistened with cleaner.

(4) Spray on red penetrant
After cleaning, spray on the red penetrant and allow 5-10 minutes for penetration. Spray on more red penetrant if it dries before it has been able to penetrate.

(5) Spray on developer
Remove any residual penetrant on the surface after the penetrant has penetrated, and spray on the surface after the penetrant has penetrated, and spray on the developer. If there are any cracks in the surface, red dots or a red line will appear several minutes after the developer dries.
Hold the developer 300-400 mm away from the area the surface uniformly.

(6) Clean the surface with the cleaner.

NOTE: Without fail, read the instructions for the color check kit before use.

5. Inspection and servicing of basic engine parts

5.1.4 Replacement of cap plugs

Step No.	Description	Procedure	Tool of material used
1	Clean and remove grease from the hole into which the cap plug is to be driven. (Remove scale and sealing material previously applied.)	Remove foreign materials with a screw driver or saw blade.	Screw driver or saw blade Thinner
2	Remove grease from the cap plug.	Visually check the nick around the plug.	Thinner
3	Apply Threebond No.4 to the seat surface where the plug is to be driven in.	Apply over the whole outside of the plug.	Threebond No.4
4	Insert the plug into the hole.	Insert the plug so that it sits correctly.	
5	Place a driving tool on the cap plug and drive it in using a hammer. 2-3 mm *Using the special tool, drive the cap plug so that the edge of the plug is 2 mm (0.0787 in) below the cylinder surface.	Drive in the plug parallel to the seating surface. 3 mm 100 mm	• Driving tool • Hammer

mm

Plug dia.	d	D
φ12	φ11.9-12.0	φ20
φ25	φ24.9-25.0	φ35
φ30	φ29.9-30.0	φ40
φ45	φ44.9-45.0	φ55

5.1.5 Cylinder bore measurement

Especially clean head surface, cylinder bores and oil holes, and check the below items after removing any carbon deposit and bonding agent.

(a) Appearance inspection
Check if there is any discoloration or crack. If crack is suspected, perform color check. Sufficiently clean the oil holes and check they are not clogged.

(b) Cylinder bore and distortion
Measure at 20 mm below the crest of the liner, at 20 mm from the bottom end and at the center in two directions A and B as shown in the below figure.

Roundness:
Roundness is found as follows though it is the simple method. Measure cylinder diameters of the A direction and the B direction on each section of a, b and c.
Roundness is the maximum value among those difference values.

Cylindricity:
Cylindricity is found as follows though it is the simple method.
Measure cylinder diameters of a, b and c sections in the A direction, and calculate the difference in maximum value and minimum value of the measured diameters.
In the same way measure and calculate the difference in the B direction.

Cylindricity is the maximum value between those difference values.

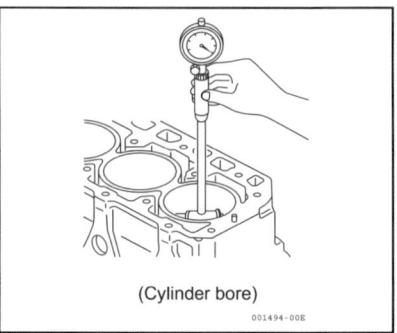
(Cylinder bore)

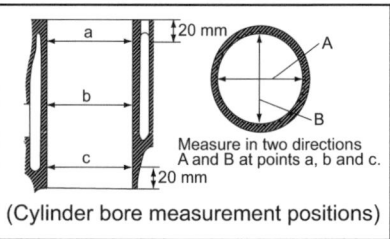
Measure in two directions A and B at points a, b and c.
(Cylinder bore measurement positions)

mm

Item	Model	Standard	Limit
Cylinder bore diameter	3YM30	76.000-76.030	76.200
	3YM20/2YM15	70.000-70.030	70.200
Cylinder roundness /Inclination		0.01 or less	0.03

5.2 Cylinder head

The cylinder head is of 3-cylinder integral construction, mounted with 14 bolts. Special alloy stellite with superior resistance to heat and wear is fitted on the seats, and the area between the valves is cooled by the water jet.

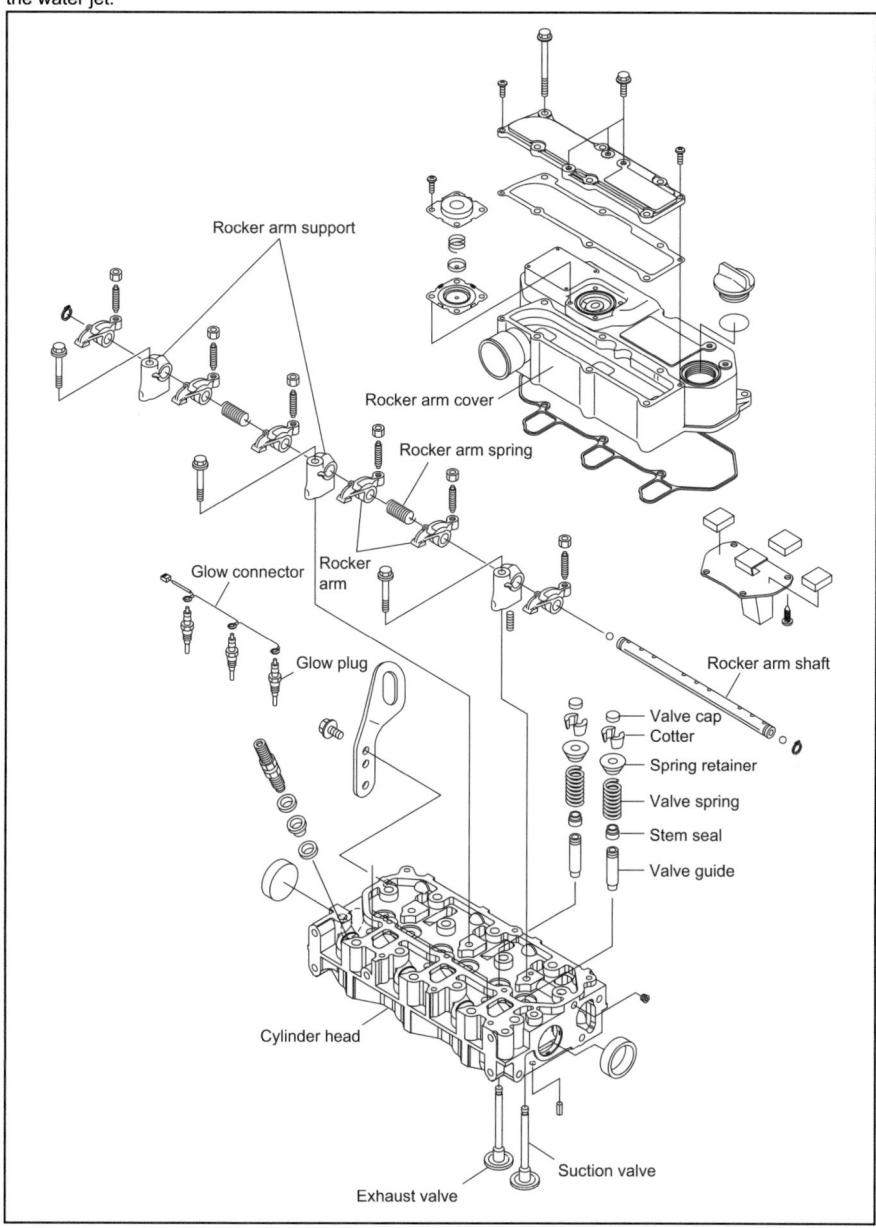

5. Inspection and servicing of basic engine parts

5.2.1 Inspecting the cylinder head

The cylinder head is subjected to very severe operating conditions with repeated high pressure, high temperature and cooling. Thoroughly remove all the carbon and dirt after disassembly and carefully inspect all parts.

(1) Distortion of the combustion surface
Carefully check for cylinder head distortion as this leads to gasket damage and compression leaks.
1) Clean the cylinder head surface.
2) Place a straight-edge along each of the four sides and each diagonal. Measure the clearance between the straight-edge and combustion surface with a feeler gauge.

mm

	Standard	Wear limit
Cylinder head distortion	0.05 or less	0.15

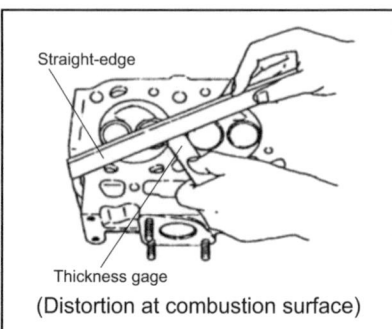

(Distortion at combustion surface)

(2) Checking for cracks in the combustion surface
Remove the fuel injection nozzle, intake and exhaust valve and clean the combustion surface. Check for discoloration or distortion and conduct a color check test to check for any cracks.

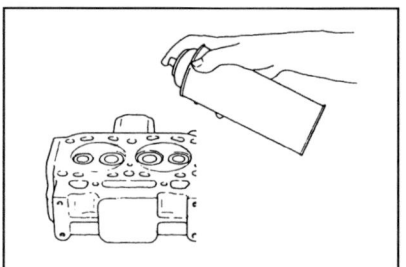

(3) Checking the intake and exhaust valve seats
Check the surface and width of the valve seats.
If they are too wide, or if the surfaces are rough, correct to the following standards:

Seat angle	Intake	120°
	Exhaust	90°

Seat width	Standard	Limit
Intake	1.07-1.24	1.74
Exhaust	1.24-1.45	1.94

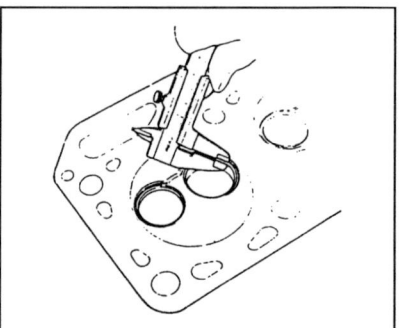

5.2.2 Valve seat correction procedure

The most common method for correcting unevenness of the seat surface with a seat grinder is as follows:

(1) Use a seat grinder to make the surface even.
As the valve seat width will be enlarged, first use a 70° grinder, then grind the seat to the standard dimension with a 15° grinder.

NOTE:
When seat adjustment is necessary, be sure to check the valve and valve guide. If the clearance exceeds the tolerance, replace the valve or the valve guide, and then grind the seat.

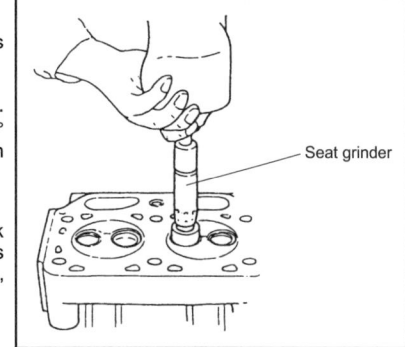

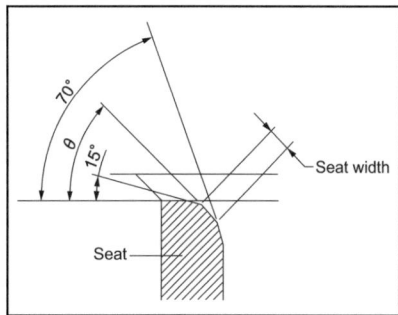

(2) Knead valve compound with oil and finish the valve seat with a lapping tool.

(3) Final finishing should be done with oil only.

Lapping tool
Use a rubber cap type lapping tool for cylinders without a lapping tool groove with oil only.

NOTE:
Clean the valve and cylinder head with light oil or the equivalent after valve seat finishing is completed and make sure that there are no grindings remaining.

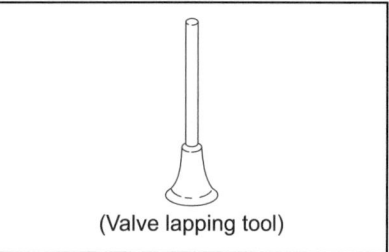
(Valve lapping tool)

NOTE:
1) Insert adjusting shims between the valve spring and cylinder head when seats have been refinishing with a seat grinder.
2) Measure valve distortion after valve seat refinishing has been completed, and replace the valve and valve seat if it exceeds the tolerance.

5.2.3 Intake/exhaust valves, valve guides

(1) Wearing and corrosion of valve stem
Replace the valve stem is excessively worn or corroded.

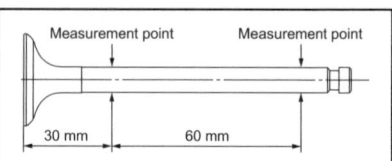

mm

Valve stem outside dia.	Standard	Limit
Intake	5.960-5.975	5.90
Exhaust	5.945-5.960	5.90

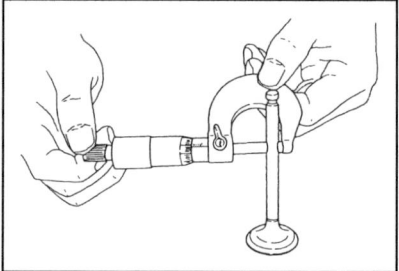

(2) Inspection of valve seat wear and contact surface
Inspect for valve seat scratches and excessive wear. Check to make sure the contact surface is normal. The seat angle must be checked and adjusted if the valve seat contact surface is much smaller than the width of the valve seat.

NOTE:
   Keep in mind the fact that the intake and discharge valve have different diameters.

(3) Valve sink
Over long periods of use and repeated lapping, combustion efficiency may drop. Measure the sinking distance and replace the valve and valve seat if the valve sink exceeds the tolerance.

mm

	Standard	Limit
Valve sink	0.4-0.6	0.8

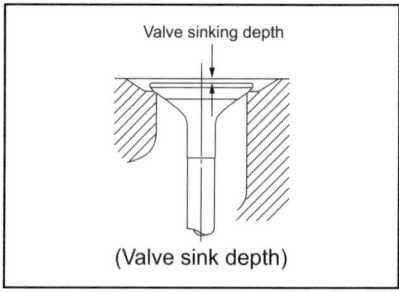

(Valve sink depth)

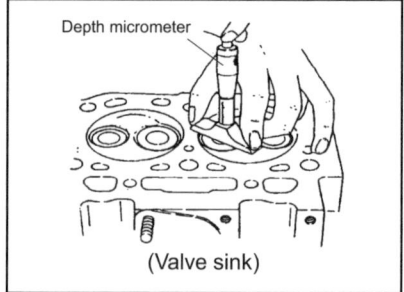

(Valve sink)

5. Inspection and servicing of basic engine parts

(4) Valve guide

1) Measuring inside diameter of valve guide.
   Measure the inside diameter of the valve guide and replace it if it exceeds the wear limit.

mm

		Standard	Limit
Valve guide inside dia.	Intake	6.000-6.012	6.08
	Exhaust	6.000-6.012	6.08
Clearance	Intake	0.025-0.052	0.16
	Exhaust	0.040-0.067	0.17

NOTE: The inside diameter standard dimensions assume a pressure fit.

2) Replacing the valve guide
   Use the insertion tool and tap in the guide with a mallet.
   The intake valve guide and exhaust valve guide are of different dimensions.

3) Valve guide projection
   a) Put liquid nitrogen or ether (or alcohol) with dry ice added in a container and put the valve guide for replacement in it for cooling. Then insert it in with a valve guide inserting tool.

⚠ CAUTION

Do not touch the cooled valve guide with bare hands to avoid skin damage.

   b) Check the inside diameter and finish to the standard inside diameter as required with a reamer.

Check the projection from the seat surface of the valve spring.

Valve guide projection (mm)
9.8-10.0

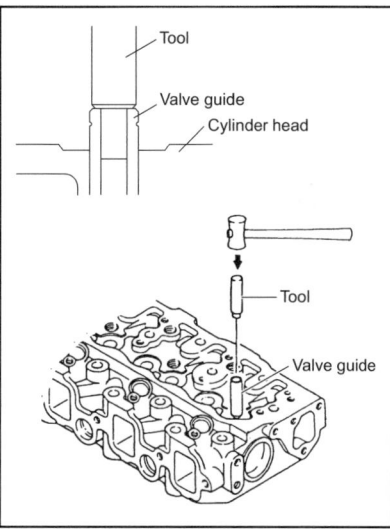

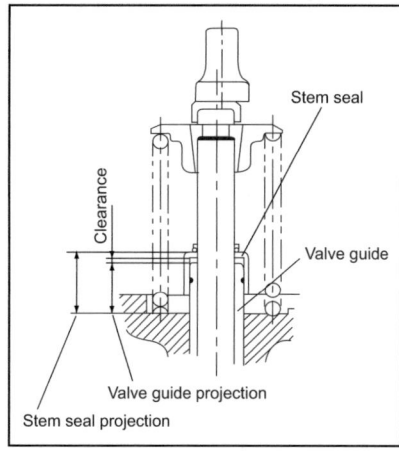

4) Valve stem seals
   The valve stem seals in the intake/exhaust valve guides cannot be re-used. When they are removed, be sure to replace them.
   When assembling the intake/exhaust valves, apply an adequate quantity of engine oil on the valve stem before inserting them.

   Exhaust stem seal is marked by yellow. Intake stem seal is not marked.

   a) Apply lube oil to the lip.
   b) Push with the inserting tool for installation.
   Measure and check the projection of valve stem seal to keep proper clearance between valve guide and stem seal.

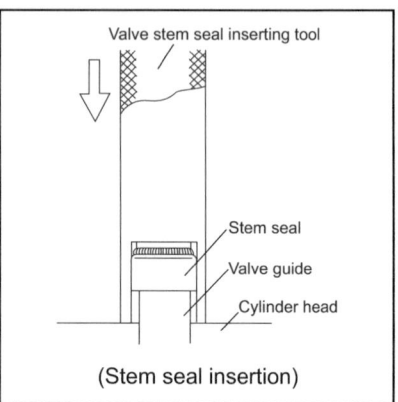

(Stem seal insertion)

Valve stem seal projection (mm)
10.9-11.2

5.2.4 Valve springs

(1) Checking valve springs

   1) Check the spring for scratches or corrosion.
   2) Measure the free length of the spring.

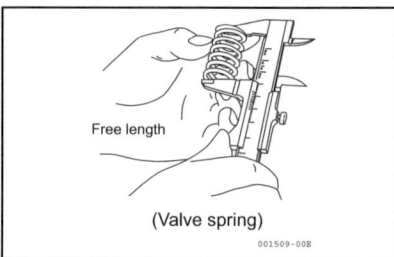

(Valve spring)

   3) Measure inclination.

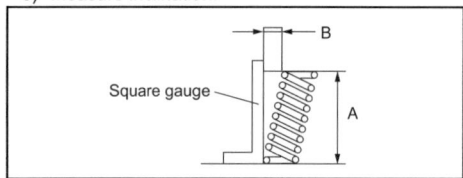

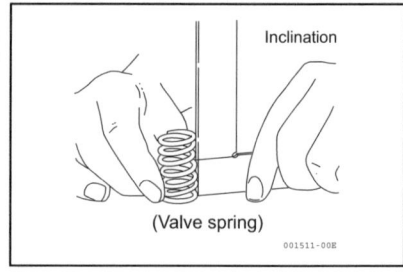

(Valve spring)

5. Inspection and servicing of basic engine parts

4) Measure spring tension.

Valve spring	Unit	Standard	Limit
Free length A	mm	37.8	36.3
Inclination B	mm	-	1.3
Tension (1 mm pressure) (Smaller pitch/ Larger pitch)	N/mm (kgf/mm)	26.6/35.4 (2.71/3.61)	-

5) Assembling valve springs.

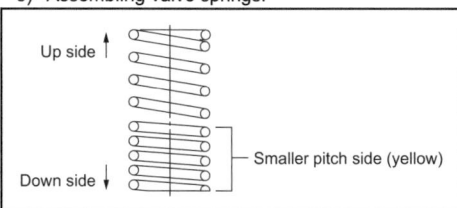

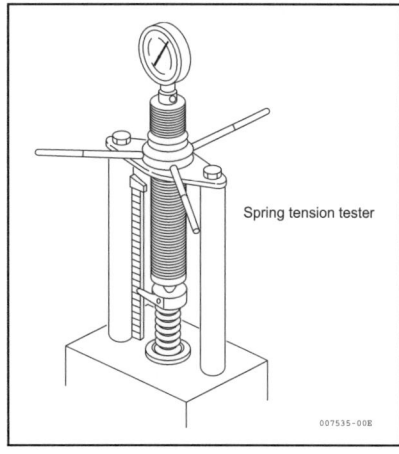

Spring tension tester

NOTE:
The pitch of the valve spring is not even. The side with the smaller pitch (yellow) should face down (cylinder head) when assembled.

6) Spring retainer and spring cotter
   Inspect the inside face of the spring retainer, the outside surface of the spring cotter, the contact area of the spring cotter inside surface and the notch in the head of the valve stem. Replace the spring retainer and spring cotter when the contact area is less than 70%, or when the spring cotter has been recessed because of wear.

5.2.5 Assembling the cylinder head

Partially tighten the bolts in the specified order and then tighten to the specified torque, being careful that the head does not get distorted.

(1) Clean out the cylinder head bolt holes.

(2) Check for foreign matter on the cylinder head surface where it comes in contact with the block.

(3) Coat the head bolt threads and nut seats with lube oil.

(4) Use the positioning pins to line up the head gasket with the cylinder block.

(5) Match up the cylinder head with the head gasket and mount.

	First	Second
	N•m (kgf•m)	
Cylinder head bolt tightening torque	27.0-33.0 (2.8-3.4)	53.9-57.9 (5.5-5.9)

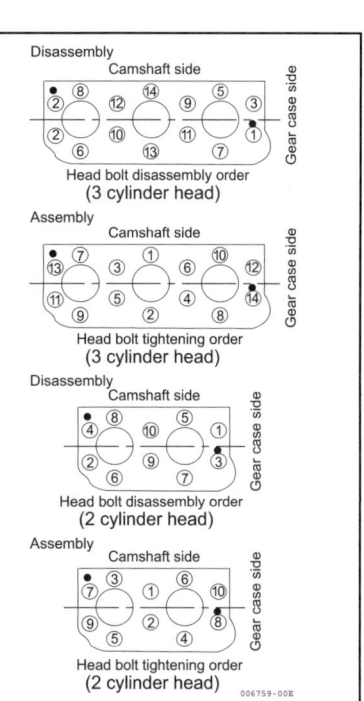

5.2.6 Measuring top clearance

(1) Place a high quality fuse (Ø1.5 mm, 10 mm long) in three positions on the flat part of the piston head.

(2) Assemble the cylinder head gasket and the cylinder head and tighten the bolts in the specified order to the specified torque.

(3) Turn the crank, (in the direction of engine revolution), and press the fuse against the piston until it breaks.

(4) Remove the head and take out the broke fuse.

(5) Measure the three positions where each fuse is broken and calculate the average.

Model	Top clearance (mm)
3YM30	0.747-0.891
3YM20/2YM15	0.697-0.841

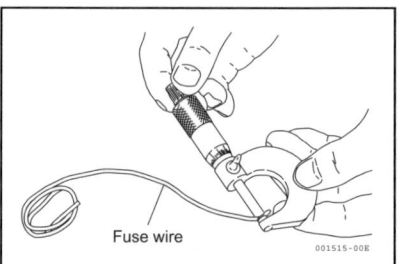

Fuse wire

5.2.7 Intake and exhaust rocker arms

The wear of rocker arm and rocker arm bushing may change opening/closing timing of the valve, and may in turn affect the engine performance according to the extent of the change.

(1) Rocker arm shaft and rocker arm bushing
Measure the outer diameter of the shaft and the inner-diameter of the bushing, and replace if wear exceeds the limit.

mm

	Standard	Limit
Outside dia of intake and exhaust rocker arm shaft.	11.966-11.984	11.94
Intake and exhaust rocker arm bushing inside dia.	12.000-12.020	12.07
Rocker arm shaft and bushing clearance at assembly	0.016-0.054	0.13

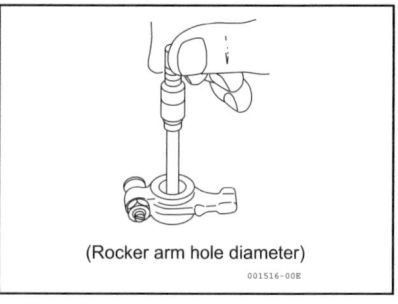

(Rocker arm hole diameter)

Replace the rocker arm bushing if it moves and replace the entire rocker arm if there is no tightening clearance.

(2) Rocker arm spring
Check the rocker arm spring and replace it if it is corroded or worn.

(3) Rocker arm and valve top retainer wear

(4) Inspect the contact surface of the rocker arm and replace it if there is abnormal wear or flaking.

5. Inspection and servicing of basic engine parts

5.2.8 Adjustment of valve clearance

(1) Make adjustments when the engine is cool.
(Refer to 2.2.2(8).)

mm

Intake and exhaust valve clearance	0.15-0.25

(2) Be sure that the opening and closing angles for both the intake and the exhaust valves are checked when the timing gear is disassembled (The gauge on the flywheel can be read.).

deg.

Intake valve Open	b.TDC.	7-17
Intake valve Closed	a.BDC.	35-45
Exhaust valve Open	b.BDC.	35-45
Exhaust valve Closed	a.TDC.	7-17

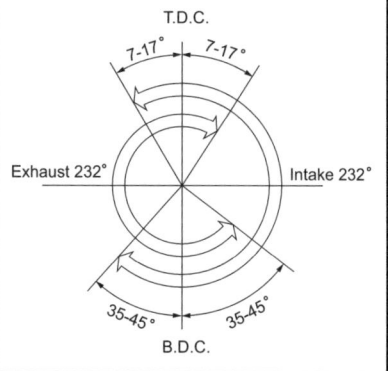

Exhaust 232°   Intake 232°

5.3 Piston and piston pins

Pistons are made of a special light alloy with superior thermal expansion characteristics, and the top of the piston forms a combustion chamber.

ID marks for piston size
ML
MS

IMPORTANT :
Piston shape differs among engine models. If any incorrect piston is installed, combustion performance will drop. Be sure to check the applicable engine model identification mark on the piston to insure use of the correct part.

Model	ID marks for engine model
3YM30	76K
3YM20/2YM15	70

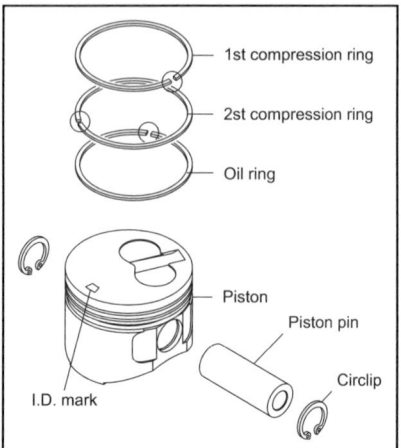

5.3.1 Piston

(1) Piston head and combustion surface
Remove the carbon that has accumulated on the piston head and combustion surface, taking care not to scratch the piston. Check the combustion surface for any damage.

(2) Measurement of piston outside diameter/ inspection
   1) Replace the piston if the outsides of the piston or ring grooves are worn.
   2) Measure the outside diameter in the position of 22mm from the piston bottom in the right angle direction of the piston pin.

Piston outside diameter

Model	Standard	Limit	Clearance between piston and cylinder
3YM30	75.965-75.975	75.920	0.035-0.055
3YM20 2YM15	69.970-69.980	69.925	0.030-0.050

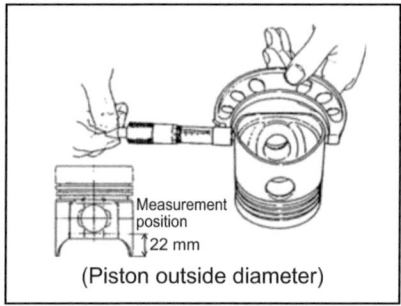

If the piston outside diameter exceed the limit, replace the piston with new one.

Selective pairing of cylinder and piston
Piston must be paired with cylinder according to the below table. The size mark of a piston is shown on the top surface of the piston and the size mark of a cylinder block is shown on the non-operating side of the cylinder block. The service parts of pistons are provided.

5. Inspection and servicing of basic engine parts

	Tolerance		Piston outside diameter. D2	
			below + 0.005 0 min.	below 0 -0.005 min.
		Size mark	ML	MS
Cylinder inside diameter D1	+0.030 max. +0.020 min.	L	O	X
	below +0.020 +0.010 min.	M	O	O
	below +0.010 0 min.	S	X	O

Model	Cylinder inside diameter D1	Piston outside diameter. D2
3YM30	76	75.970
3YM20/2YM15	70	69.975

(3) Removing the piston pin
A floating type piston pin is used in this engine. The piston pin can be pressed into the piston pin hole at room temperature (Coat with oil to make it slide easily).

5.3.2 Piston pin

Measure the outer diameter and replace the pin if it is excessively worn.

	Standard	Limit
Piston pin hole inside dia.	22.000-22.009	22.039
Piston pin outside dia.	21.995-22.000	21.965
Clearance	0.000-0.014	0.074

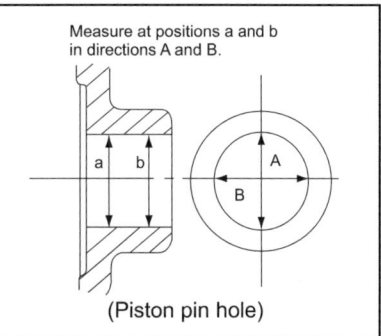

Measure at positions a and b in directions A and B.

(Piston pin hole)

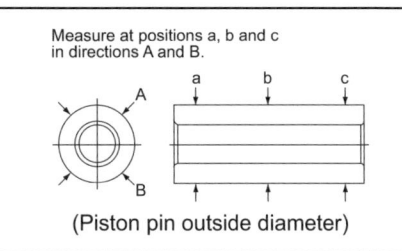

Measure at positions a, b and c in directions A and B.

(Piston pin outside diameter)

5.3.3 Piston rings

There are two compression rings and one oil ring.
The absence of an oil ring on the piston skirt prevents oil from being kept on the thrust surface and in turn provides good lubrication.

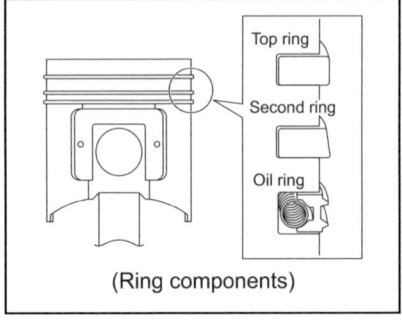

(Ring components)

(1) Measuring the rings.

Measure the thickness and width of the rings, and the ring-to-groove clearance after installation. Replace if wear exceed the limit.

3YM30   mm

		Standard	Limit
Top ring	Groove width	1.550-1.570	-
	Ring width	1.470-1.490	1.450
	Clearance	0.060-0.100	-
Second ring	Groove width	1.580-1.595	1.695
	Ring width	1.430-1.450	1.410
	Clearance	0.013-0.165	0.285
Oil ring	Groove width	3.010-3.030	3.130
	Ring width	2.970-2.990	2.950
	Clearance	0.020-0.060	0.180

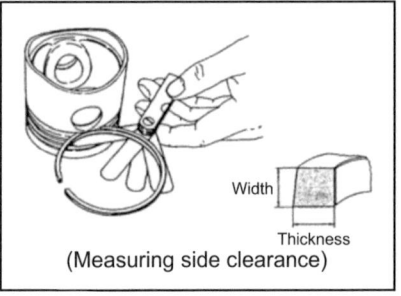

(Measuring side clearance)

3YM20/2YM15   mm

		Standard	Limit
Top ring	Groove width	1.550-1.570	-
	Ring width	1.470-1.490	1.450
	Clearance	0.060-0.100	-
Second ring	Groove width	1.540-1.560	1.660
	Ring width	1.470-1.490	1.450
	Clearance	0.050-0.090	0.210
Oil ring	Groove width	3.010-3.030	3.130
	Ring width	2.970-3.010	2.950
	Clearance	0.020-0.060	0.180

5. Inspection and servicing of basic engine parts

(2) Measuring piston ring gap

Press the piston ring onto a piston liner and measure the piston ring gap with a gauge. Press on the ring about 30mm from the bottom of the liner.

3YM30/3YM20/2YM15   mm

		Standard	Limit
Top ring gap		0.15-0.30	0.390
Second ring gap		0.18-0.33	0.420
Oil ring gap	3YM30	0.20-0.45	0.540
	3YM20 2YM15	0.15-0.35	0.44

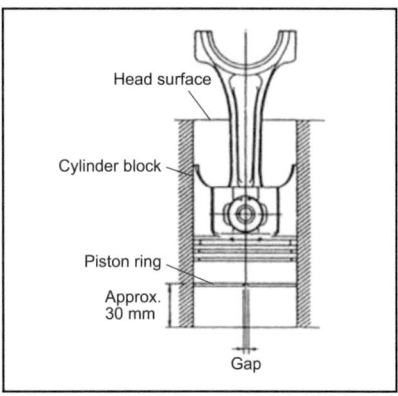

(3) Removing the piston rings

1) Thoroughly clean the ring grooves when removing piston rings.
2) The side with the manufacturer's mark should face up.

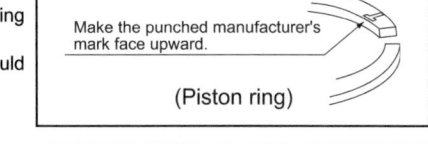

3) After fitting the piston ring, make sure it moves easily and smoothly.
4) Stagger the piston rings at 120° intervals, making sure none of them line up with the piston.

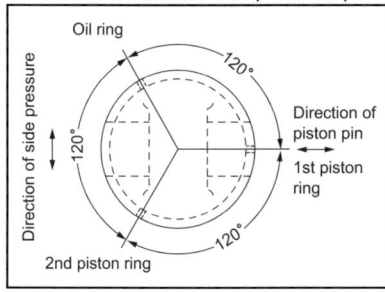

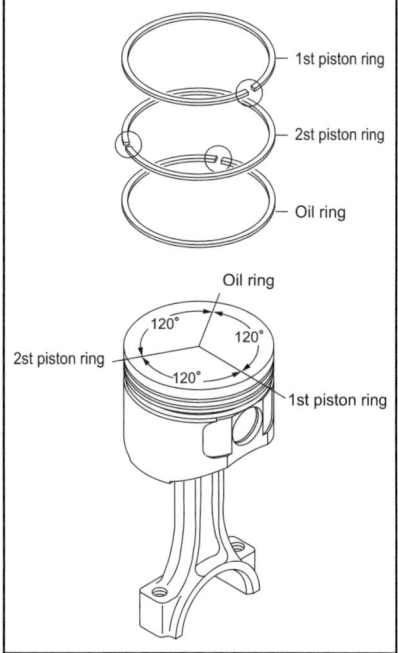

119R1

5) The oil ring is provided with a coil expander. The coil expander joint should be opposite (staggered 180°) the oil ring gap.

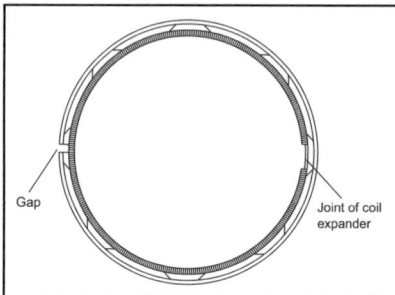

5.4 Connecting rod

The connecting rod is made of high-strength forged carbon steel.
The large end with the aluminium metal can be separated into two and the small end has a 2-layer copper alloy coil bushing.

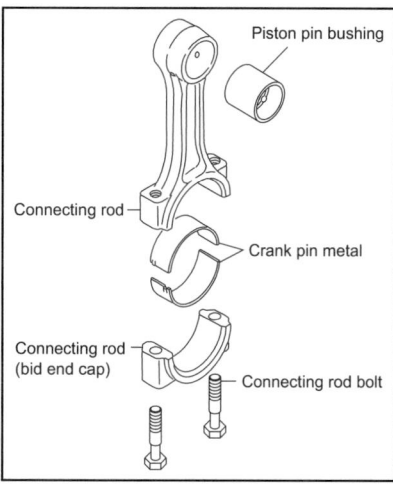

5.4.1 Inspecting the connection rod

(1) Twist and parallelism of the large and small ends

Insert the measuring tool into the large and small ends of the connecting rod. Measure the extent of twist and parallelism and replace if they exceed the tolerance.

mm

	Standard	Limit
Connecting rod twist and parallelism	Less than 0.03 per 100 mm	0.08

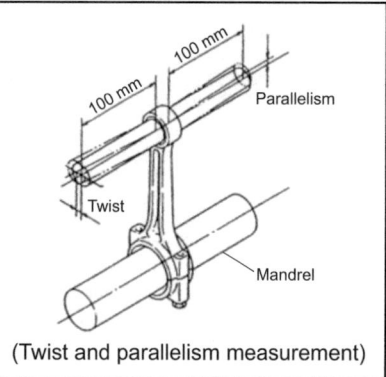

(Twist and parallelism measurement)

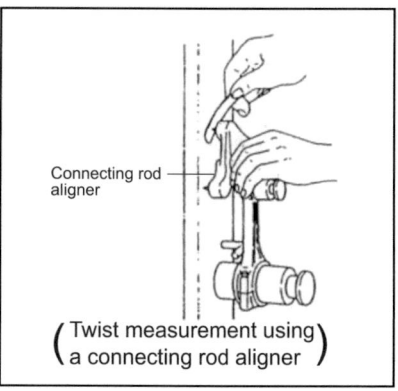

(Twist measurement using a connecting rod aligner)

5. Inspection and servicing of basic engine parts

(2) Checking the side clearance of a connecting rod

Fit the respective crank pins to the connecting rod and check to make sure that the side clearance in the crankshaft direction is correct.

	Standard	Limit
Connecting rod side clearance	0.20-0.40	0.55

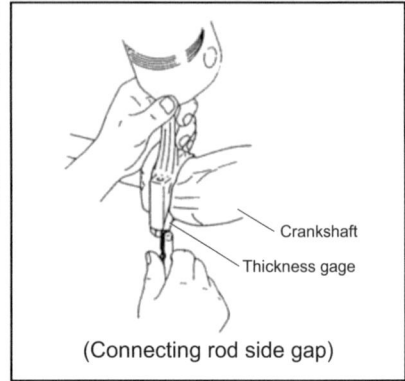

(Connecting rod side gap)

5.4.2 Crank pin metal

(1) Checking crank pin metal

Check for flaking, melting or seizure on the contact surface of the crank pin metal.

(2) Measuring crank pin oil clearance

Measure the crankpin outside diameter and the crank pin metal inside diameter. Calculate the oil clearance from the measured values.
(Refer to 5.5.1(3) for measuring the crank pin outside diameter.)

Replace the crank pin metal if the oil clearance becomes about the limit dimension of the below table.
Correct by grinding if unevenly wear, roundness exceeding the limit or insufficient outside diameter is found. Also use an undersized metal if necessary.

[NOTICE]
When measuring the inside diameter of the rod big end, install the crank pin metal in the rod big end not to mistake the top and bottom of the metals and tighten the rod bolts by the standard torque.

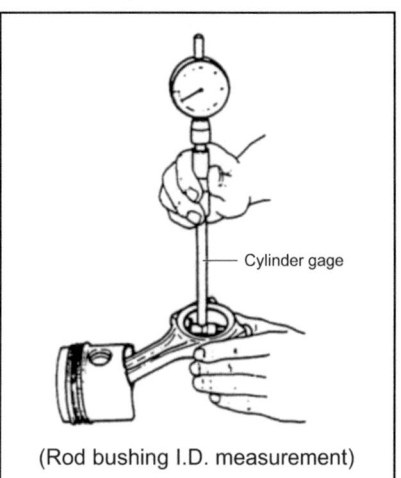

	N·m (kgf·m)
Rod bolt tightening torque	22.6-27.5 (2.3-2.8)

(Rod bushing I.D. measurement)

3YM30   mm

Item	Standard	Limit
Crankpin O.D.	41.952-41.962	41.902
Metal I.D.	41.982-42.010	-
Metal thickness	1.503-1.509	-
Clearance	0.020-0.058	0.120

3YM20/2YM15   mm

Item	Standard	Limit
Crankpin O.D.	37.952-37.962	37.902
Metal I.D.	37.982-38.010	-
Metal thickness	1.503-1.509	-
Clearance	0.020-0.058	0.120

5. Inspection and servicing of basic engine parts

- Other procedure of measuring crank pin oil clearance
  1) Use the press gauge (Plastic gauge) for measuring the oil clearance of the crank pin.
  2) Mount the connecting rod on the crank pin (tighten to specified torque).
  3) Remove the connecting rod and measure the broken plastic gauge with measuring paper.

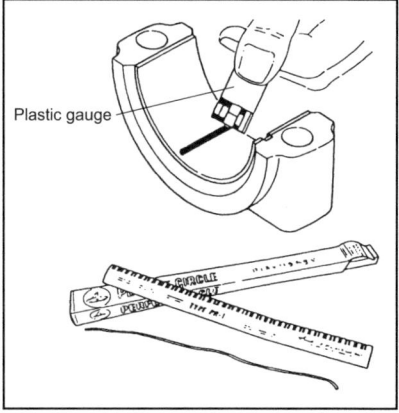
Plastic gauge

(3) Precautions on replacement of crank pin metal
  1) Wash the crank pin metal.
  2) Wash the large end cap, mount the crank pin metal and make sure that it fits tightly on the large end cap.
  3) When assembling the connecting rod, match up the large end and large end cap number. Coat the bolts with engine oil and gradually tighten them alternately to the specified torque.
     If a torque wrench is not available, make match marks on the bolt heads and large end cap before disassembling (to indicate the proper torque position) and retighten the bolts to those positions.
  4) Make sure there is no sand, metal cuttings or other foreign matter in the lube oil, and that the crankshaft is not scratched. Take special care in cleaning the oil holes.

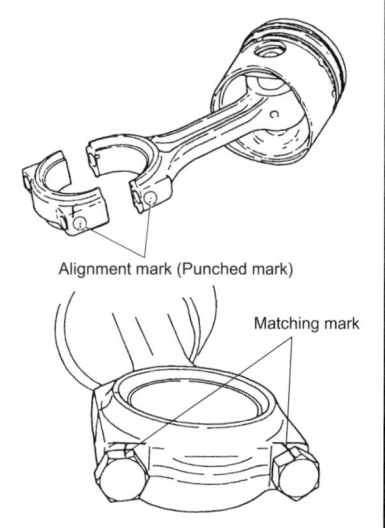
Alignment mark (Punched mark)
Matching mark

5.4.3 Piston pin bushing

(1) Measuring piston pin clearance.
Excessive piston pin bushing wear may result in damage to the piston pin or the piston itself.
Measure the piston pin bushing inside diameter and the piston pin outside diameter. Calculate the oil clearance from the measured values. (Refer to 5.3.2 for the piston pin.)

mm

	Standard	Limit
Piston pin metal I.D.	22.025-22.038	22.068
Pin O.D.	21.991-22.000	21.963
Clearance	0.025-0.047	0.105

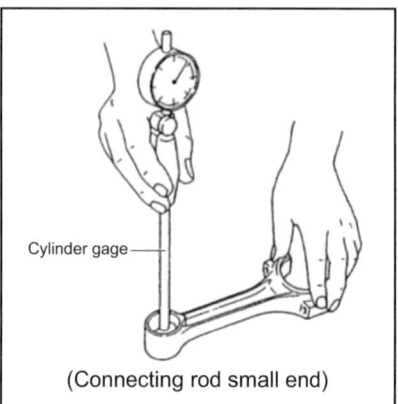

(Connecting rod small end)

5.4.4 Assembling piston and connecting rod

The piston and connecting rod should be assembled so that the match mark on the connecting rod large end faced the fuel nozzle side and the combustion chamber above the piston is close to the fuel nozzle.

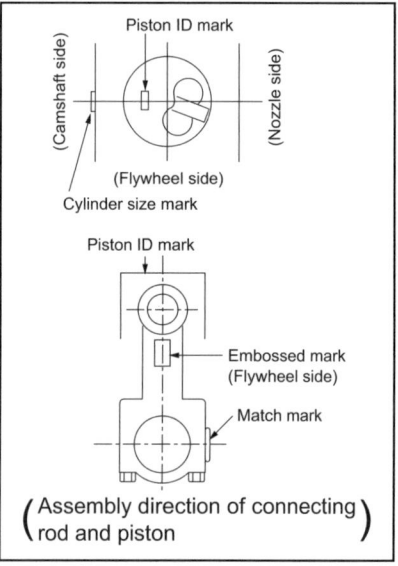

$\left(\begin{array}{c}\text{Assembly direction of connecting}\\ \text{rod and piston}\end{array}\right)$

5. Inspection and servicing of basic engine parts

5.5 Crankshaft and main bearing

The crank pin and crank journal have been induction hardened for superior durability, and the crankshaft is provided with four balance weights for optional balance. The crankshaft main bearing is of the hanger type. The upper metal (cylinder block side) is provided with an oil groove. There is no oil groove on the lower metal (bearing cap side). The bearing cap (location cap) of the flywheel side has a thrust metal which supports the thrust load.

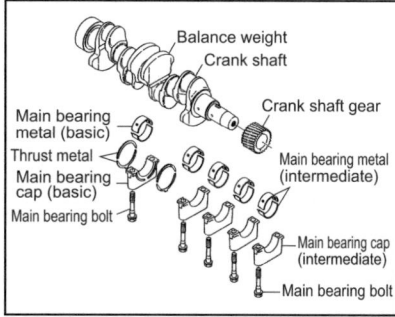

5.5.1 Crankshaft

(1) Color check after cleaning the crankshaft, and replace the crank shaft if there is any cracking or considerable damage.

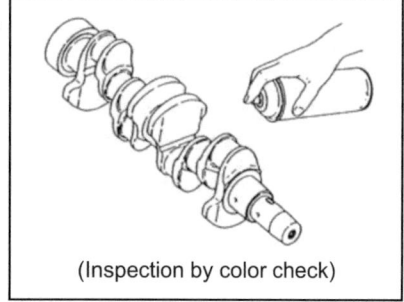

(Inspection by color check)

(2) Bending of the crankshaft

Support the crankshaft with V-blocks at both ends of the journals. Measure the deflection of the center journal with a dial gauge while rotating the crankshaft to check the extent of crankshaft bending. The total indicating reading on the dial gauge is divided by two to obtain the crankshaft bend.

mm

Crankshaft bend limit	0.01

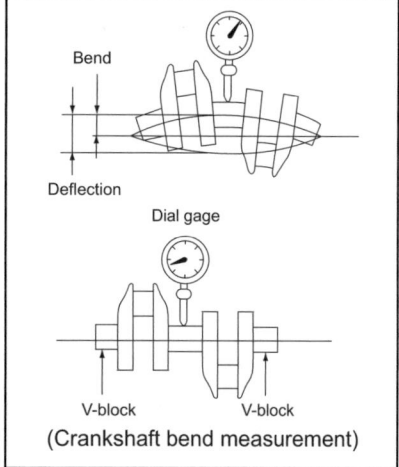

(Crankshaft bend measurement)

5. Inspection and servicing of basic engine parts

(3) Measuring the crank pin and journal

Measure the outside diameter, roundness and taper at each crank pin and journal.
Correct by grinding if unevenly wear, roundness exceeding the limit or insufficient outside diameter is found. Replace if the defect is excessive.

mm

Item	Standard (Diameter)	Limit (Diameter)
Roundness Taper	0.01 or less	0.02

To look for the oil clearance of crank pin, measure the inside diameter of crank pin metal. (Refer to 5.4.2(2).)

mm

		Standard	Limit
Crank pin 3YM30	Outside dia.	41.952-41.962	41.902
	Oil clearance	0.020-0.058	0.120
Crank pin 3YM20/2YM15	Outside dia.	37.952-37.962	37.902
	Oil clearance	0.020-0.058	0.120
Crank journal (Selective pair) All models	Outside dia.	46.952-46.962	46.902
	Oil clearance	0.020-0.050	0.120

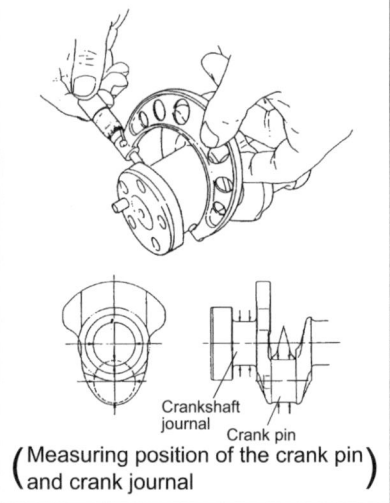

(Measuring position of the crank pin and crank journal)

- Dimension R and finishing precision of crankshaft journal and pin

As for grinding processing of journal and pin, machine it by using the grinding wheel of the dimension R of below table.

Surface finishing precision standard on journal and pin:

Ry = 0.8S super polishing

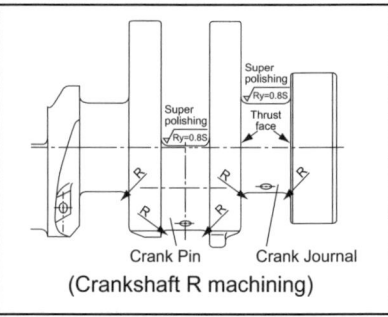

(Crankshaft R machining)

Surface finishing precision standard on the thrust side of crankshaft arm :

Finishing precision standard of dimension R (mm)
3.5+0.3/0

NOTICE:
1) If the oil clearance is excessive though the thickness of the journal and crankpin metals are normal or if partial uneven wear is observed, re-grind the crankshaft and use an undersize metals.
2) If rust or surface roughening exists on the rear side of the metals, coat it with blue or minimum. Then assemble the crankpin metal to the connecting rod, and tighten the rod bolt to the specified torque to check the metal for contact. If the contact surface occupies 75% or more, the metal is normal. If the contact surface is insufficient, the metal interference is insufficient. Replace the metal with a new one.

5. Inspection and servicing of basic engine parts

(4) Checking the side gap of a crankshaft

After assembling the crankshaft, tighten the main bearing cap to the specified toque, and move the crankshaft to one side, placing a dial gauge on one end of the shaft to measure thrust clearance.

Replace the thrust bearing if it is worn beyond the limit.

mm

	Standard	Limit
Crankshaft side gap	0.111-0.250	0.30

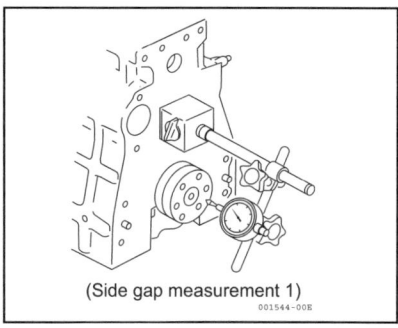

(Side gap measurement 1)

Other measurement method can also be effective. Insert the thickness gauge directly into the clearance between the thrust metal and crankshaft thrust face.

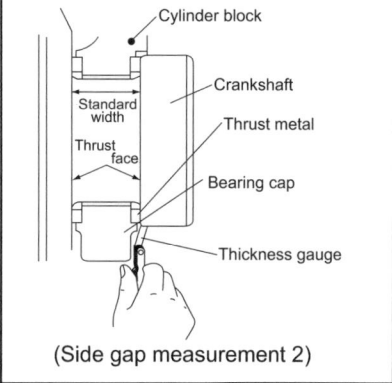

(Side gap measurement 2)

5.5.2 Main bearing

(1) Inspecting the main bearing
Check for flaking, seizure or burning of the contact surface and replace if necessary.

(2) Measuring the inner diameter of metal
Tighten the cap to the specified torque and measure the inner diameter of the metal.

N·m (kgf·m)

Tightening torque	75.5-81.5 (7.7-8.3)

NOTE:
When assembling the bearing cap, keep the following in mind.
1) The lower metal (cap side) has no oil groove.
2) The upper metal (block side) has an oil groove.
3) Check the cylinder block alignment number.
4) The "FW" on the cap lies on the flywheel side.

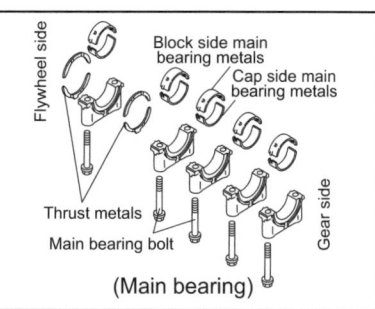

(Main bearing)

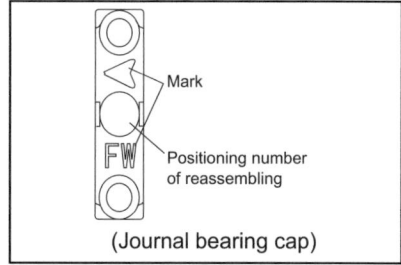

(Journal bearing cap)

5.6 Camshaft and tappets

5.6.1 Camshaft

The camshaft is normalized and the cam and bearing surfaces are surface hardened and ground. The cams have a curve that minimized the repeated shocks on the valve seats and maximizes valve seat life.

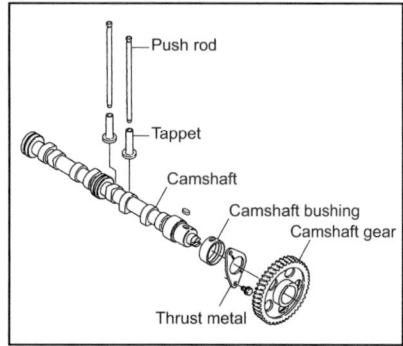

(1) Checking the camshaft side gap

Measure the thrust gap before disassembly. As the cam gear is shrink-fitted to the cam, be careful when replacing the thrust bearing.

mm

	Standard	Limit
Camshaft side gap	0.05-0.15	0.25

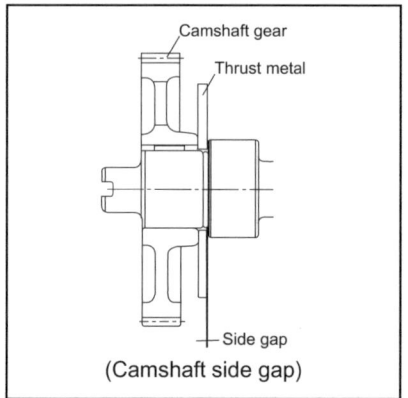

(Camshaft side gap)

(2) Measure the cam height, and replace the cam if it is worn beyond the limit.

3YM30   mm

		Standard	Limit
Cam height	Intake	34.135-34.265	33.89
	Exhaust		

3YM20/2YM15   mm

		Standard	Limit
Cam height	Intake	34.535-34.665	34.29
	Exhaust		

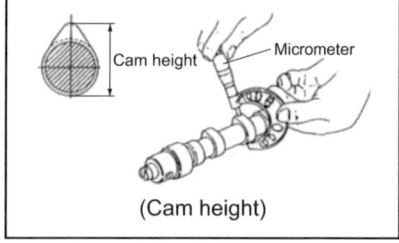

(Cam height)

5. Inspection and servicing of basic engine parts

(3) Camshaft and bearing hole measurement

mm

		Standard	Limit
Cam height	Intake	34.135-34.265	33.89
	Exhaust		

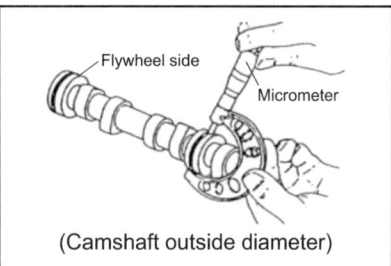

(Camshaft outside diameter)

Measure the camshaft outside diameter with a micrometer. The oil clearance shall be calculated by subtracting the measured camshaft outside diameter from the inside diameter of the camshaft bearing or bushing. The camshaft bushing at gear case side is measured with a cylinder gage after insertion to the cylinder. Replace if they exceed the limit or are damaged.

mm

	Item	Standard	Limit
Gear side	Bushing I.D.	40.000-40.075	40.150
	Camshaft O.D.	39.940-39.960	39.905
	Oil clearance	0.040-0.135	0.245
Intermediate position	Bushing I.D.	40.000-40.025	40.100
	Camshaft O.D.	39.910-39.935	39.875
	Oil clearance	0.065-0.115	0.225
Flywheel side	Bushing I.D.	40.000-40.025	40.100
	Camshaft O.D.	39.940-39.960	39.905
	Oil clearance	0.040-0.085	0.195

(4) Bending of the camshaft

Support both ends of the camshaft with V-blocks, place a dial gauge at the central bearing areas and measure bending. Replace if excessive.

NOTE:
The reading on the dial gauge is divided by two to obtain the camshaft bend.

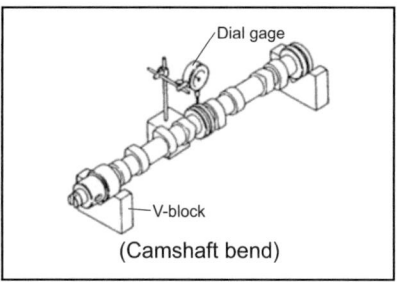

(Camshaft bend)

mm

	Standard	Limit
Camshaft bend	0.02 or less	0.05

5.6.2 Tappets

(1) The tappets are offset to rotate during operation and thereby prevent uneven wearing. Check the contact of each tappet and replace if excessively or unevenly worn.

NOTE:
When removing tappets, be sure to keep them separate for each cylinder and intake/exhaust valve.

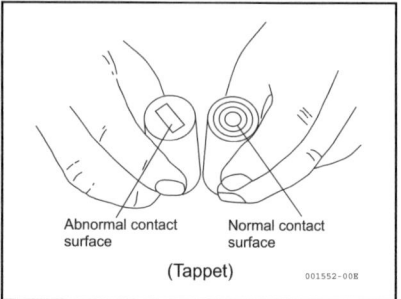

Abnormal contact surface   Normal contact surface

(Tappet)

(2) Measure the outer diameter of the tappet, and replace if worn beyond the limit.

mm

	Standard	Limit
Tappet outside diameter	20.927-20.960	20.907
Tappet guide hole inside dia. (cylinder block)	21.000-21.021	21.041
Oil clearance	0.040-0.094	0.134

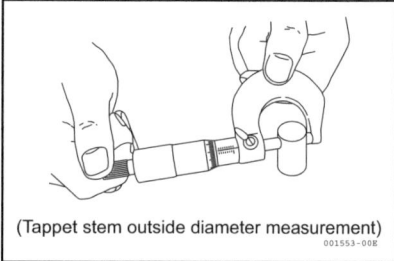

(Tappet stem outside diameter measurement)

(3) Measuring push rods.
Measure the bend of the push rods.

mm

	Standard	Limit
Push rod bend	Less than 0.03	0.03

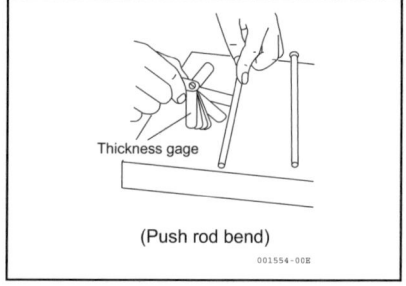

Thickness gage

(Push rod bend)

5.7 Timing gear

The timing gear is helical type for minimum noise and specially treated for high durability.

5.7.1 Inspecting the gears

(1) Inspect the gears and replace if the teeth are damaged or worn.

(2) Measure the backlash of all gears that mesh, and replace the meshing gears as a set if wear exceeds the limit.

NOTE:
If backlash is excessive, it will not only result in excessive noise and gear damage, but also lead to bad valve and fuel injection timing and a decrease in engine performance.

mm

	Standard	Limit
Backlash	0.06-0.12	0.14

(3) Idling gear
The bushing is pressure fitted into the idling gear.
Measure the bushing inner diameter and the outer diameter of the shaft, and replace the bushing or idling gear shaft if the oil clearance exceeds the wear limit.
A, B and C are inscribed on the end of the idling gear. When assembling, these marks should align with those on the cylinder block.

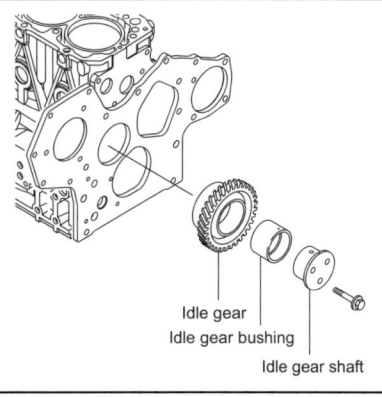

mm

	Standard	Limit
Idle gear shaft diameter	36.950-36.975	36.900
Idle gear bushing inside dia.	37.000-37.025	37.075
Oil clearance	0.025-0.075	0.175

5.7.2 Gear timing marks

Match up the timing marks on each gear when assembling (A, B and C).

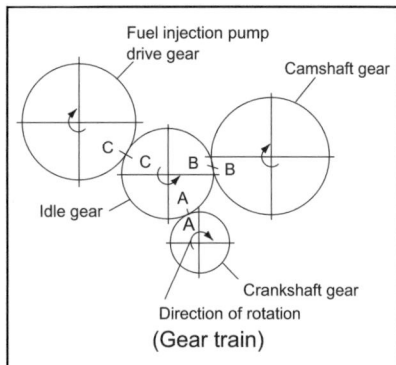

(Gear train)

5. Inspection and servicing of basic engine parts

5.8 Flywheel and housing

The function of the flywheel is through inertia, to rotate the crankshaft in a uniform and smooth manner by absorbing the turning force created during the combustion stroke of the engine, and by compensating for the decrease in turning force during the other strokes.
The flywheel is mounted and secured by 6 bolts on the crankshaft end at the opposite end to the gear case; it is covered by the mounting flange (flywheel housing) which is bolted to the cylinder block.
The fitting surface for the damper disc is on the crankshaft side of the flywheel. The rotation of the crankshaft is transmitted through this disc to the input shaft of the reduction and reversing gear. The reduction and reversing gear is fitted to the mounting flange.
The flywheel's unbalanced force on the shaft center must be kept below the specified value for the crankshaft as the flywheel rotates with the crankshaft at high speed.
To achieve this, the valance is adjusted by drilling holes in the side of the flywheel, and the unbalanced momentum is adjusted by drilling holes in the circumference.
The ring gear is shrink fitted onto the circumference of the flywheel, and this ring gear serves to start the engine by meshing with the starter motor pinion.
The stamped letter and line which show top dead center of each cylinder are positioned on the flywheel circumference, and by matching these marks with the arrow mark at the hole of the flywheel housing, the rotary position of the crankshaft can be ascertained in order to adjust tappet clearance or fuel injection timing.

5.8.1 Position of top dead center and fuel injection timing

(1) Marking

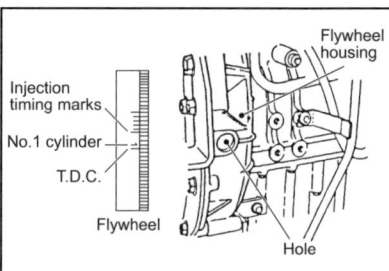

(2) Matching mark
The matching mark is made at the hole of the flywheel housing.

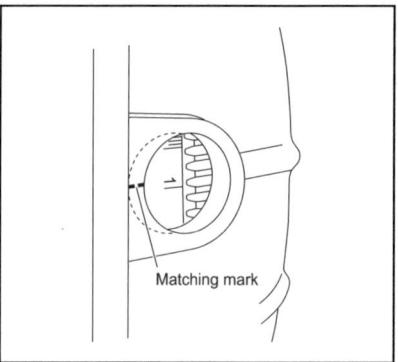

5.8.2 Damper disc

(1) Spline part
Whenever uneven wear and/or scratches are found, replace with new one.

(2) Spring
Whenever uneven wear and/or scratches are found, replace with new one.

(3) Pin wear
Whenever uneven wear and/or scratches are found, replace with new one.

(4) Whenever a crack or damage to the spring slot is found, replace the defective part with new one.

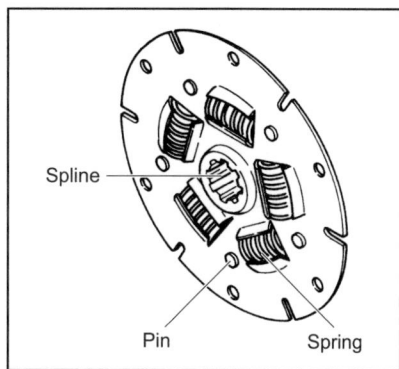

6. Fuel injection equipment

6.1 Fuel Injection pump/governor

Refer to the service manual of the YPES-ML fuel injection pump for the disassembly, assembly and adjustment procedure.

6.1.1 Fuel system diagram

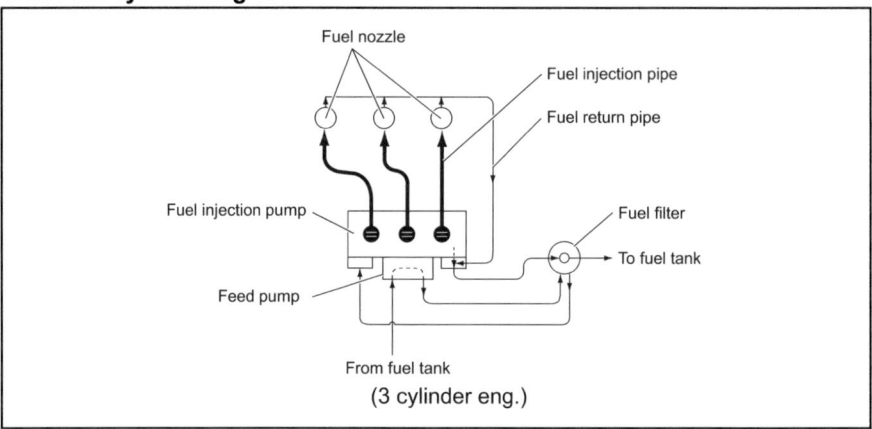

(3 cylinder eng.)

6. Fuel injection equipment

6.1.2 Fuel injection pump service data and adjustment

(1) Service data
  1) Service data for 3YM30

Part code (Back No.)			-	\multicolumn{2}{c}{728990-51350 (TMR1)}
Adjustment SPEC			-	ENGINE SPEC	SERVICE STD
Item	Fuel valve (Valve pressure)		-	A (120)	(170)
	Nozzle type (ID mark)		-	NP-DN0PDN158	DN-12SD12
	Fuel injection pipe		mm	Ø 2 x 320	Ø2 x 600
Injection adjustment	Starting	Pump speed N_D	min^{-1}	200	-
		Average injection volume Q_D	mm^3/st	38 ± 5	-
	Rated load	Pump speed N_A	min^{-1}	1800	-
		Injection volume Q_A	mm^3/st	22.4 ± 0.75	-
		Variation	%	± 3	-
	Torque rise	Pump speed N_C	min^{-1}	1300	-
		Injection volume Q_C	mm^3/st	25.9 ± 0.5	-
		Variation	%	± 5	-
	Hi-idle	Pump speed N_B	min^{-1}	1925	-
		Injection volume Q_B	mm^3/st	(4)	-
	Idle	Pump speed N_E	min^{-1}	425	-
		Injection volume Q_E	mm^3/st	(4)	-
Plunger stroke			mm	\multicolumn{2}{c}{7.0}
Plunger diameter			mm	\multicolumn{2}{c}{Ø 6}
Retraction volume of delivery valve			mm	\multicolumn{2}{c}{20.1}
Pre-stroke			mm	\multicolumn{2}{c}{2.9}
Top clearance			mm	\multicolumn{2}{c}{1.4}
Governor spring	Spring constant		N/cm (kgf/cm)	\multicolumn{2}{c}{4.23 (0.431)}
	Free length		mm	\multicolumn{2}{c}{42}

Note : The value in parentheses is a reference value.

6. Fuel injection equipment

2) Service data for 3YM20

Part code (Back No.)			-	728890-51300 (TMM1)	
Adjustment SPEC			-	ENGINE SPEC	SERVICE STD
Item	Fuel valve (Valve pressure)		-	C (120)	(170)
	Nozzle type (ID mark)		-	NP-DN0PDN159	DN-12SD12
	Fuel injection pipe		mm	Ø 2 x 300	Ø2 x 600
Injection adjustment	Starting	Pump speed N_D	min^{-1}	200	-
		Average injection volume Q_D	mm^3/st	30± 5	-
	Rated load	Pump speed N_A	min^{-1}	1800	-
		Injection volume Q_A	mm^3/st	17.4 ± 0.5	-
		Variation	%	± 3	-
	Torque rise	Pump speed N_C	min^{-1}	1300	-
		Injection volume Q_C	mm^3/st	19.6 ± 0.5	-
		Variation	%	± 5	-
	Hi-idle	Pump speed N_B	min^{-1}	1945	-
		Injection volume Q_B	mm^3/st	(6)	-
	Idle	Pump speed N_E	min^{-1}	425	-
		Injection volume Q_E	mm^3/st	(5)	-
Plunger stroke			mm	7.0	
Plunger diameter			mm	Ø 6	
Retraction volume of delivery valve			mm	23.5	
Pre-stroke			mm	2.5	
Top clearance			mm	1.0	
Governor spring	Spring constant		N/cm (kgf/cm)	5.54 (0.565)	
	Free length		mm	42	

Note : The value in parentheses is a reference value.

6. Fuel injection equipment

3) Service data for 2YM15

Part code (Back No.)			-	\multicolumn{2}{c}{728890-51300 (TMM1)}	
Adjustment SPEC			-	ENGINE SPEC	SERVICE STD
Item	Fuel valve (Valve pressure)		-	C (120)	(170)
	Nozzle type (ID mark)		-	NP-DN0PDN159	DN-12SD12
	Fuel injection pipe		mm	Ø 2 x 300	Ø2 x 600
Injection adjustment	Starting	Pump speed N_D	min^{-1}	200	-
		Average injection volume Q_D	mm^3/st	30± 5	-
	Rated load	Pump speed N_A	min^{-1}	1800	-
		Injection volume Q_A	mm^3/st	17.9 ± 0.5	-
		Variation	%	± 3	-
	Torque rise	Pump speed N_C	min^{-1}	1300	-
		Injection volume Q_C	mm^3/st	20.6 ± 0.5	-
		Variation	%	± 5	-
	Hi-idle	Pump speed N_B	min^{-1}	1925	-
		Injection volume Q_B	mm^3/st	(6)	-
	Idle	Pump speed N_E	min^{-1}	425	-
		Injection volume Q_E	mm^3/st	(5)	-
Plunger stroke			mm	\multicolumn{2}{c}{7.0}	
Plunger diameter			mm	\multicolumn{2}{c}{Ø 6}	
Retraction volume of delivery valve			mm	\multicolumn{2}{c}{23.5}	
Pre-stroke			mm	\multicolumn{2}{c}{2.5}	
Top clearance			mm	\multicolumn{2}{c}{1.0}	
Governor spring		Spring constant	N/cm (kgf/cm)	\multicolumn{2}{c}{5.54 (0.565)}	
		Free length	mm	\multicolumn{2}{c}{42}	

Note : The value in parentheses is a reference value.

6. Fuel injection equipment

(2) Fuel adjustment procedure
- Fuel adjustment for 3YM30 and 2YM15

1) Loosen the nut of the FO limiter, and screw the adjusting bolt to the bottom. Tighten the nut.
2) Adjust the rack position to R0 (15mm) with the FO limiter, and measure the injection volume (A) at the rated speed.
3) When the measured injection volume (A) is out of the standard, readjust the injection volume by screwing the FO limiter.
4) Pull the control lever to high idle side at full rack position.
5) Adjust the position of the high idle set bolt at high idle speed as to get the specified injection volume (B).
6) Loosen the nut of the FO limiter, and loosen the adjusting bolt at 360°. Fix the nut.
7) Screw the angleich at the torque-rise speed to get the specified injection volume (C).
8) Confirm the injection volume (D) at the specified speed at starting.
9) Push the control lever to low idle side.
10) Adjust the position of the low idle set bolt at low idle speed as to get the specified injection volume (E).

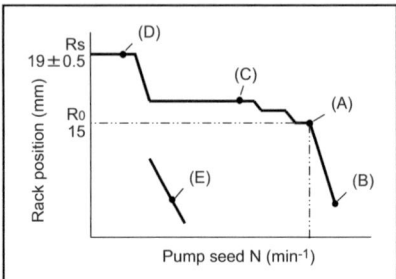

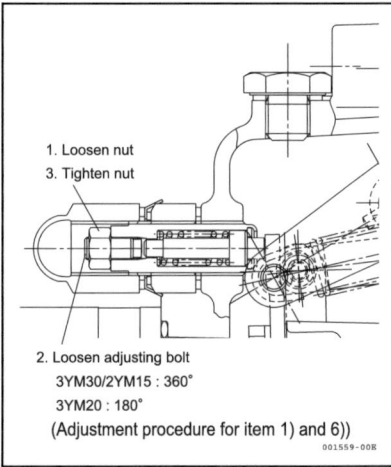

1. Loosen nut
3. Tighten nut

2. Loosen adjusting bolt
3YM30/2YM15 : 360°
3YM20 : 180°
(Adjustment procedure for item 1) and 6))

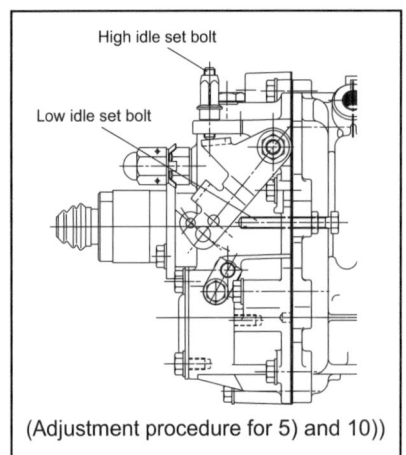

High idle set bolt
Low idle set bolt

(Adjustment procedure for 5) and 10))

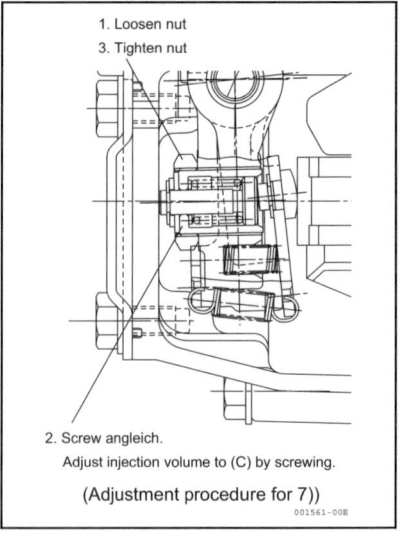

1. Loosen nut
3. Tighten nut

2. Screw angleich.
Adjust injection volume to (C) by screwing.

(Adjustment procedure for 7))

6. Fuel injection equipment

- Fuel adjustment for 3YM20
  1) Loosen the nut of the FO limiter, and screw the adjusting bolt to the bottom. Tighten the nut.
  2) Adjust the rack position to R0 (15mm) with the FO limiter, and measure the injection volume (A) at the rated speed.
  3) When the measured injection volume (A) is out of the standard, readjust the injection volume by screwing the FO limiter.
  4) Pull the control lever to high idle side at full rack position.
  5) Adjust the position of the high idle set bolt at high idle speed as to get the specified injection volume (B).
  6) Loosen the nut of the FO limiter, and loosen the adjusting bolt at 180°. Fix the nut.
  7) Screw the angleichi at the torque-rise speed to get the specified injection volume (C).
  8) Confirm the injection volume (D) at the specified speed at starting.
  9) Push the control lever to low idle side.
  10) Adjust the position of the low idle set bolt at low idle speed as to get the specified injection volume (E).

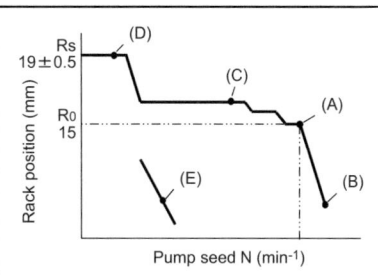

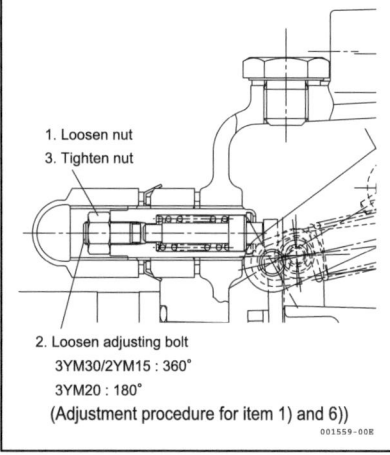

1. Loosen nut
3. Tighten nut

2. Loosen adjusting bolt
   3YM30/2YM15 : 360°
   3YM20 : 180°
(Adjustment procedure for item 1) and 6))

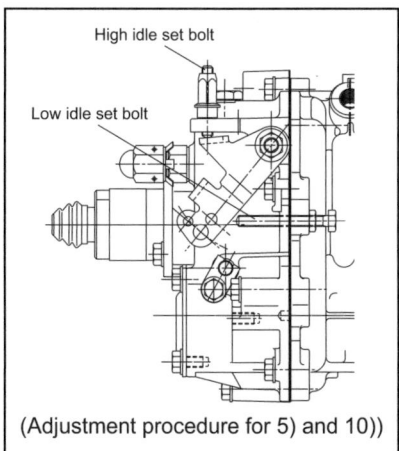

(Adjustment procedure for 5) and 10))

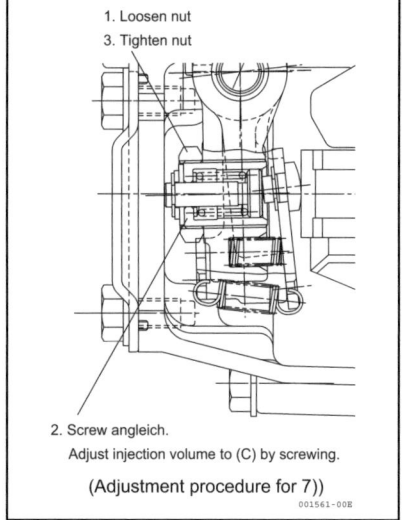

1. Loosen nut
3. Tighten nut

2. Screw angleich.
Adjust injection volume to (C) by screwing.
(Adjustment procedure for 7))

138R1

6.1.3 Fuel injection pump structure

Section of a fuel injection pump/ governor for 3YM30.

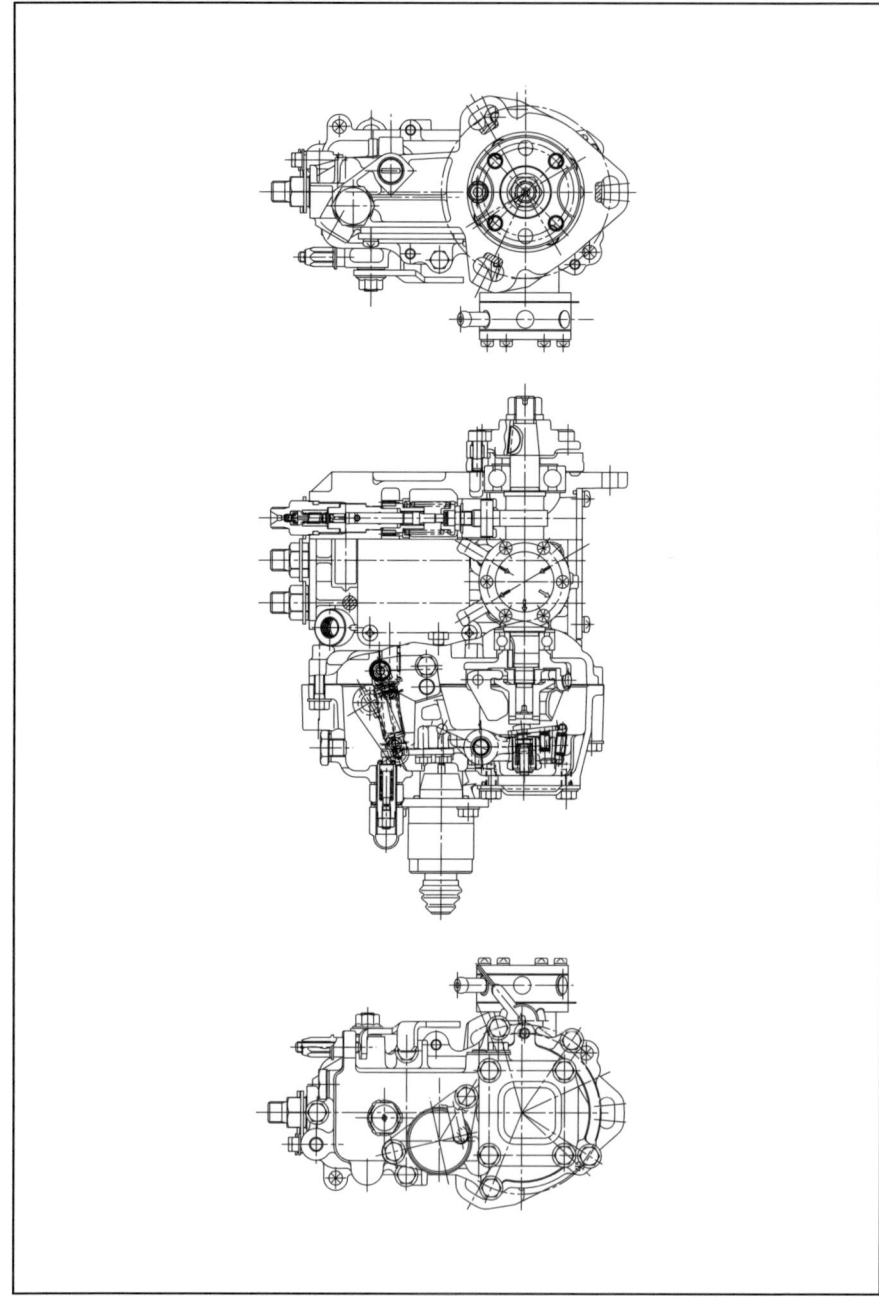

6.1.4 Removing a fuel injection pump

The procedure to remove a fuel injection pump from the gear case is shown.

[NOTICE]
Be sure to remove a flange and a fuel injection pump drive gear with a pair without loosening the flange bolts.

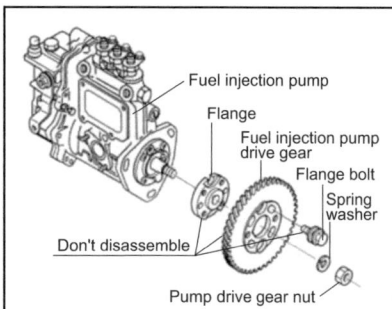

1) Remove fuel injection pipes, fuel pipes and a remote control wire. Block the entrance with tape so that trash may not enter the fuel injection pipes and the fuel injection pump.
2) Mark the position of the timing marks of a fuel pump and a gear case. Or, put a mark on the gear case at the position to agree the timing mark of a fuel pump.
3) Remove a pump cover from the gear case.
4) Give the marks on a fuel injection pump drive gear and a idle gear with paint or the like.
5) Loosen fuel injection pump installation nuts (three nuts).
6) Loosen a installation nut of a fuel injection pump drive gear.
7) Pull a fuel injection pump drive gear and a flange with a pair to your side by gear puller.
8) Remove a pump drive gear nut and a washer.
9) Remove a fuel injection pump. Leave the pump drive gear in the gear case.

6.1.5 Installing a fuel injection pump

[NOTICE]
- Use a new O-ring on the fuel pump flange and apply grease.
- Confirm whether the marks of the pump drive gear and the idle gear is correct.

1) Turn a camshaft so that the key of the pump camshaft may almost agree in a position of the key groove of a pump drive gear.
2) Insert a fuel injection pump into the installation hole of the gear case straight to prevent the damage of the O-ring. Insert a fuel pump while confirming whether the key of a camshaft and the key groove of a drive gear agree.
3) Assemble a pump drive gear nut and a washer together temporarily.
4) Turn a fuel injection pump to the position where the marks of the fuel injection pump and the gear case agrees.
5) Fasten pump installation nuts (three nuts) on the fuel injection pump.
6) Tighten the pump installation nut by the specified standard torque.

Tightening torque of the pump drive gear nut (without lube oil)

Tighteningtorque	N•m(kgf•m)
58.8-68.8 (6.0-7.0)	

6.1.6 Adjusting fuel injection timing

Refer to 2.2.6(1) in chapter 2.

6. Fuel injection equipment

6.1.7 Troubleshooting of fuel injection pump

Complete repair means not only replacing defective parts, but finding and eliminating the cause of the trouble as well. The cause of the trouble may not necessarily be in the pump itself, but may be in the engine or the fuel system. If the pump is removed prematurely, the true cause of the trouble may never be known. Before removing the pump from the engine, at least go through the basic check points given here.

Basic checkpoints
- Check for breaks or oil leaks throughout the fuel system, from the fuel tank to the nozzle.
- Check the injection timings for all cylinders. Are they correctly adjusted? Are they too fast or too slow?
- Check the nozzle spray.
- Check the fuel delivery. Is it in good condition? Loosen the fuel pipe connection at the injection pump inlet, and test operate the fuel feed pump.

6.1.8 Major faults and troubleshooting

Fault		Cause	Remedy
1. Engine won't start	Fuel not delivered to injection pump	(1) No fuel in the fuel tank. (2) Fuel tank cock is closed. (3) Fuel pipe system is clogged (4) Fuel filter element is clogged (5) Air is sucked into the fuel due to defective connections in the piping from the fuel tank to the fuel pump. (6) Defective valve contact of feed pump (7) Piston spring of feed pump is broken. (8) Inter-spindle or tappets of feed pump are stuck	Re-supply Open Clean Disassemble and clean, or replace element Repair Repair or replace Replace Repair or replace
	Fuel delivered to injection pump.	(1) Defective connection of control lever and accel rod of injection pump. (2) Plunger is worn out or stuck. (3) Delivery valve is stuck (4) Control rack doesn't move (5) Injection pump coupling is damaged, or the key is broken.	Repair or adjust Repair or replace Repair or replace Repair or replace Replace
	Nozzle doesn't work.	(1) Nozzle valve doesn't open or close normally (2) Nozzle seat is defective. (3) Case nut is loose. (4) Injection nozzle starting pressure is too low (5) Nozzle spring is broken. (6) Fuel oil filter is clogged. (7) Excessive oil leaks from the nozzle sliding area.	Repair or replace Repair or replace Inspect and tighten Adjust Replace Repair or replace Replace the nozzle assembly
	Injection timing is defective.	(1) Injection timing is retarded due to failure of the coupling. (2) Camshaft is excessively worn. (3) Roller guide incorrectly adjusted or excessively worn. (4) Plunger is excessively worn.	Adjust Replace camshaft Adjust or replace Replace plunger assembly

6. Fuel injection equipment

Fault		Cause	Remedy
2. Engine starts, but immediately stops.		(1) Fuel pipe is clogged.	Clean
		(2) Fuel filter is clogged.	Disassemble and clean, or replace the element
		(3) Improper air- tightness of the fuel pipe connection or pipe is broken and air is being sucked in.	Replace packing repair pipe
		(4) Insufficient fuel delivery from the feed pump	Repair or replace
3. Engine's output is insufficient.	Defective injection timing and other failures.	(1) Knocking sounds caused by improper (too fast) injection timing.	Inspect and adjust
		(2) Engine overheats or emits large amount of smoke due to improper (to slow) injection timing.	Inspect and adjust
		(3) Insufficient fuel delivery from feed pump.	Repair or replace
	Nozzle movements is defective.	(1) Case nut loose.	Inspect and retighten
		(2) Defective injection nozzle performance.	Repair or replace nozzle
		(3) Nozzle spring is broken.	Replace
		(4) Excessive oil leaks from nozzle.	Replace nozzle assembly
	Injection pump is defective.	(1) Max delivery limit boll is screwed in too far.	Adjust
		(2) Plunger is worn	Replace
		(3) Injection amount is not uniform.	Adjust
		(4) Injection timings are not even.	Adjust
		(5) The 1st and 2nd levers of the governor and the control rack or the injection pump are improperly lined up.	Repair
		(6) Delivery stopper is loose.	Inspect and retighten
		(7) Delivery packing is defective.	Replace packing
		(8) Delivery valve seat is defective.	Repair or replace
		(9) Delivery spring is broken.	Replace
4. Idling is rough.		(1) Movement of control rack is defective.	
		1) Stiff plunger movement or sticking.	Repair or replace
		2) Rack and pinion fitting is defective.	Repair
		3) Movement of governor is improper	Repair
		4) Delivery stopper is too tight.	Inspect and adjust
		(2) Uneven injection volume.	adjust
		(3) Injection timing is defective.	adjust
		(4) Plunger is worn and fuel injection adjustment is difficult.	Replace
		(5) Governor spring is too weak.	Replace
		(6) Feed pump can't feed oil at low speeds.	Repair or replace
		(7) Fuel supply is insufficient at low speeds due to clogging of fuel filter.	Disassemble and clean, or replace element
5. Engine runs at high speeds, but cuts out at low speeds.		(1) The wire or rod of the accel is caught.	Inspect and repair
		(2) Control rack is caught and can't be moved.	Inspect and repair
6. Engine doesn't reach max. speed		(1) Governor spring is broken or excessively worn.	Replace
		(2) Injection performance or nozzle is poor	Repair or replace

6. Fuel injection equipment

Fault		Cause	Remedy
7. Loud knocking.		(1) Injection timing is too fast or too slow. (2) Injection from nozzle is improper. Fuel drips after each injection. (3) Injection nozzle starting pressure is too high (4) Uneven injection. (5) Engine overheats, or insufficient compression.	Adjust Adjust Adjust Adjust Repair
8. Engine exhausts too much smoke.	When exhaust smoke is black:	(1) Injection timing is too fast. (2) Air volume intake is insufficient. (3) The amount of injection is uneven. (4) Injection from nozzle is improper.	Adjust Inspect and repair Adjust Repair or replace
	When exhaust smoke is white:	(1) Injection timing is too slow. (2) Water is mixed in fuel. (3) Shortage or lube oil in the engine. (4) Engine is over-cooled.	Adjust Inspect fuel system, and clean Repair Inspect

6.2 Fuel feed pump

The fuel feed pump feeds fuel from the fuel tank, passes it through the fuel filter element, and supplies it to the fuel injection pump.

The fuel feed pump is mounted on the side of this engine and is driven by the (eccentric) cam of the fuel pump camshaft. It is provided with a manual priming lever so that fuel can be supplied when the engine is stopped.

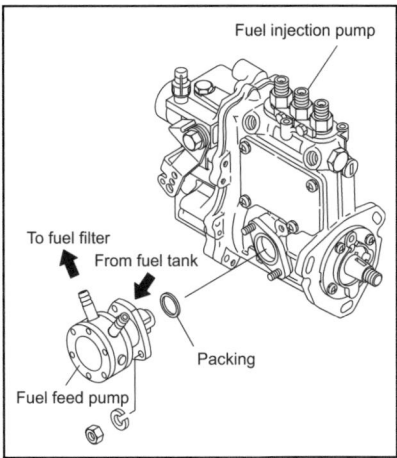

6.2.1 Construction of fuel feed pump

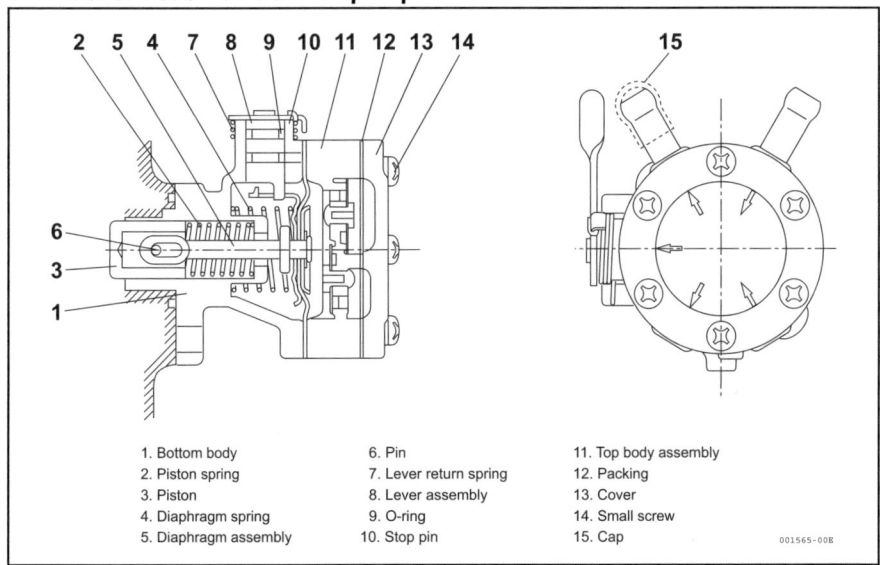

1. Bottom body
2. Piston spring
3. Piston
4. Diaphragm spring
5. Diaphragm assembly
6. Pin
7. Lever return spring
8. Lever assembly
9. O-ring
10. Stop pin
11. Top body assembly
12. Packing
13. Cover
14. Small screw
15. Cap

6.2.2 Fuel feed pump specifications

Head	1m
Discharge volume	230 cm^3/min at 1500 min^{-1} (cam), discharge pressure of 0.020 MPa (0.2 kgf/cm^2)
Closed off pressure	0.029 MPa (0.3 kgf/cm^2) or more at 400 min^{-1} (cam)

6.2.3 Disassembly and reassembly of fuel feed pump

(1) Disassembly
1) Remove the fuel feed pump mounting nut, and take the fuel feed pump off the fuel injection pump.
2) Clean the fuel feed pump assembly with fuel oil.
3) After checking the orientation of the arrow on the cover, make match marks on the upper body and cover, remove the small screw, and disassemble the cover, upper body and lower body.

(2) Reassembly
1) Clean all parts with fuel oil, inspect, and replace any defective parts.
2) Replace any packing on parts that have been disassembled.
3) Make sure that the intake valve and discharge valve on upper body are mounted in the proper direction , and that you don't forget the valve packing.
4) Assemble the diaphragm into the body, making sure the diaphragm mounting holes are lined up (do not force).
5) Align the match marks on the upper body of the pump and cover, and tighten the small screws evenly.

	N•cm(kgf•cm)
Tightening torque	147-245 (15-25)

6.2.4 Fuel feed pump inspection

(1) Place the fuel feed pump in kerosene, cover the discharge port with your finger, move the priming lever and check for air bubbles (Repair or replace any part which emits air bubbles).

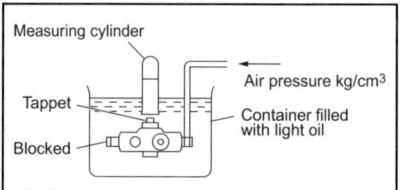

(2) Attach a vinyl house to the fuel feed pump intake, keep the pump at the specified depth from the fuel oil surface, move the priming lever by hand and check for sudden spurts of fuel oil from the discharge port. If oil is not spurted out, inspect the diaphragm and diaphragm spring and repair/replace as necessary.

(3) Diaphragm inspection
Parts of the diaphragm that are repeatedly burned will become thinner or deteriorate over a long period of time.
Check the diaphragm and replace if necessary.

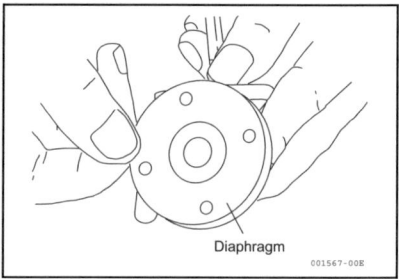

6. Fuel injection equipment

(4) Valve contact/mounting
Clean the valve seat and valve with air to remove any foreign matter.

(5) Inspect the diaphragm spring and piston spring for settling and the piston for wear, and replace as necessary.

NOTE: Replace parts as an assembly.

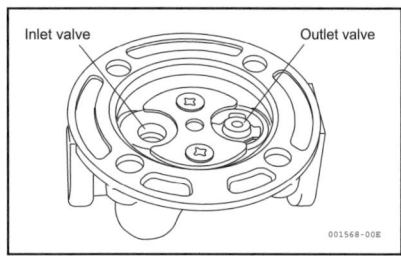

6.3 Fuel filter

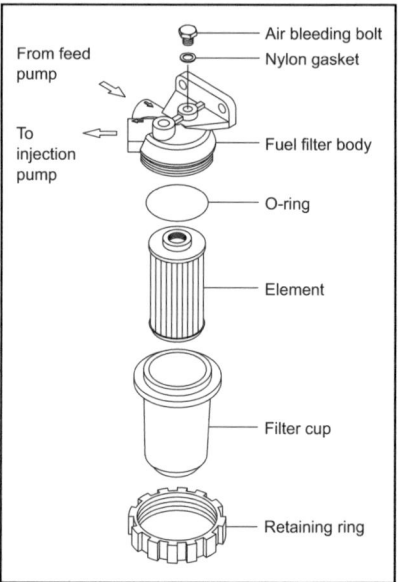

The fuel filter is installed between the feed pump and injection pump, and serves to remove dirt and impurity from the fuel fed from the fuel tank through the feed pump.

The fuel filter incorporates a replaceable filter paper element.

Fuel from the fuel tank enters the outside of the element and passes through the element under its own pressure.

As it passes through, the dirt and impurity in the fuel are filtered out, allowing only clean fuel to enter the interior of the element.

The fuel exits from the outlet at the top center of the filter and is sent to the injection pump.

Loosen the bolt fitted to the fuel filter body before starting or after dismantling and reassembly to bleed the air in the fuel system to the fuel filter.

6.3.1 Fuel filter specifications

Filtering area	333 cm^2 (20.3 in.^{2})
Material of element	Cotton fiber
Filter mesh	10-15 μ

6.3.2 Fuel filter inspection

The fuel filter must be periodically inspected. If there is water and sediment in the filter, remove all dirt, rust, etc. by washing the filter with clean fuel.

The normal replacement interval for the element is 250 hours, but the element should be replaced whenever it is dirty or damaged, even if the 250 hour replacement period has not elapsed.

6. Fuel injection equipment

6.4 Fuel tank

A triangular 30 liter fuel tank with a 2000mm (78.7402 in.) rubber fuel hose to fit all models is available as an option.
A fuel return connection is provided on top of the tank of which a rubber hose can be connected to return fuel from the fuel nozzles.

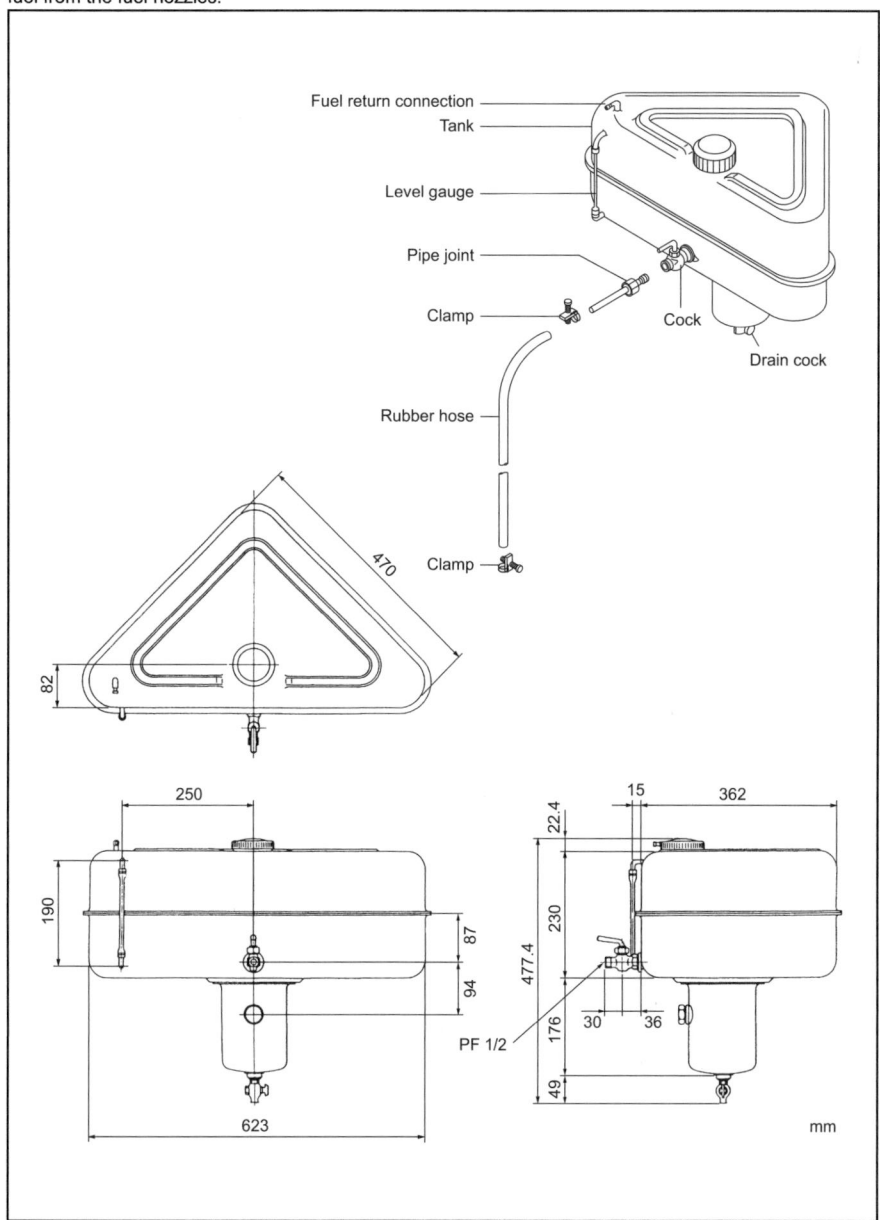

7. Intake and exhaust system

7.1 Intake system

Air enters in the intake silencer mounted at the end of the intake manifold (rocker arm cover). It is fed to the intake manifold and then on to each cylinder.
Exhaust gas goes into exhaust manifold (in the fresh water tank) mounted on the cylinder head outlet. After cooling, the exhaust gas enters the mixing elbow, which is directly connected with the exhaust manifold, and is discharged from the ship along with seawater.
When the inside of the intake manifold becomes dirty, intake air resistance is increased and reduces engine power. Periodically check the inside of the intake manifold. In the same way, the intake air silencer should be checked for dirt periodically and cleaned.
Care should also be taken to insure there is no air leakage.
Do not operate with the intake air silencer removed.

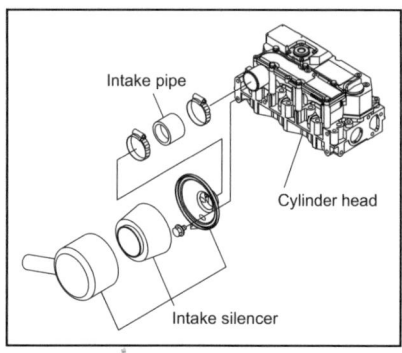

7.1.1 Breather system (A reductor to intake air system of blowby gas)

Emitting blowby gas is harmful to natural environment. Therefore blowby gas reductor is adopted as breather system.
Some of the combustion gas passes through the clearance between cylinder and piston, and flows to the crankcase. This is said as blowby gas. While it passes into cylinder head and rocker arm cover, the blowby gas mixes with splash oil, and becomes oil mist-blowby gas. The gas passes through the baffle plate inside a rocker arm cover. And it passes through a diaphragm assy, and reaches a intake silencer. The gas is reduced in the combustion chamber.
Pressure inside a crankcase is controlled by the function of the diaphragm assy, and suitable amount of blowby gas is reduced in intake air system.

[Disassemble]
When a rocker arm cover is taken off, check whether oil or the like enter the diaphragm space from a small hole on the side of a diaphragm cover or not without disassembling the diaphragm.

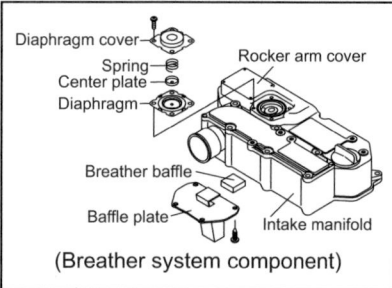

(Breather system component)

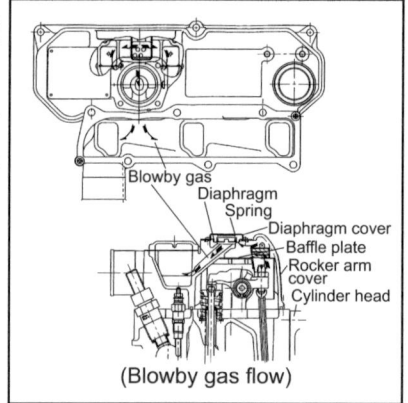

(Blowby gas flow)

[NOTICE]
1) When a diaphragm is damaged, pressure control inside the crankcase becomes insufficient, and troubles occur. When the internal pressure of the crankcase decreases too much due to the damage of a spring, much blowby gas containing oil is reduced in intake air system, and it may cause the combustion defect by the early dirt of the intake valve or the urgent rotation of the engine by the oil burning.
When pressure progresses in the crank case too much due to the wrong operation of the diaphragm and so on, it is considered that oil leakage from the joint of a oil pan, a oil seal and so on will occur. When a diaphragm is damaged, blowby is discharged from the breathing hole on the side of diaphragm cover, and not reduced in the intake manifold. Therefore, be careful of the diaphragm trouble.
2) At lubricating oil replacement or lube oil supply
The amount of lubricating oil isn't to be beyond the standard upper limit (in the engine horizontality, the upper limit mark of the dipstick). Since the blowby gas reductor is adopted, be careful that the amount of oil mist may be inducted in the combustion chamber and the oil hammer sometimes may occur, when the lubricating oil quantity is beyond the upper limit or an engine is operated beyond the allowable maximum angle of an engine.

[Reassembly]
Replace the diaphragm with new one, when it is damaged.

7.1.2 Diaphragm assy inspection

Refer to 2.2.6(5) for the inspection procedure.

7. Intake and exhaust system

7.2 Exhaust system

7.2.1 Construction

There are two types of mixing elbows, the L-type and the U-type. The mixing elbow is attached to the exhaust manifold.

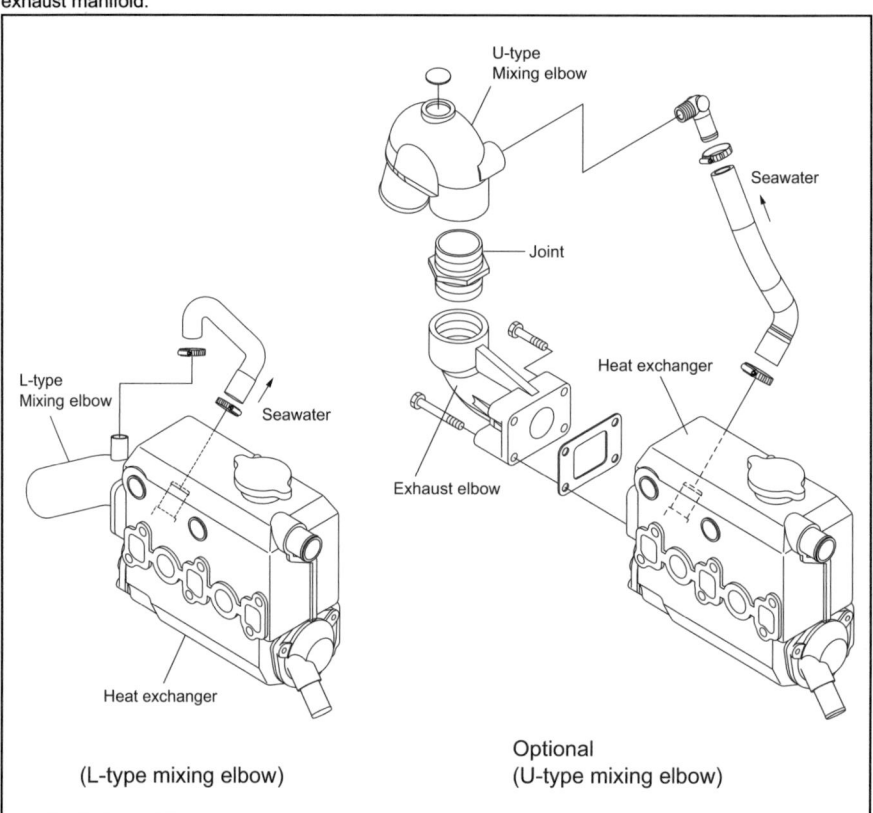

(L-type mixing elbow)

Optional
(U-type mixing elbow)

7.2.2 Mixing elbow inspection

(1) Clean dirt and scale out of the gas and cooling water lines.

(2) Repair crack or damage to welds, or replace.

(3) Inspect the gasket packing and replace as necessary.

8. Lubrication system

8.1 Lubrication system

The lube oil in the oil pan is pumped up through the intake filter and intake piping by the lube oil pump, through the holes in the cylinder body and on to the lube oil pump, through the holes in the cylinder body and on to the discharge filter.
The lube oil, which flows from the holes in the cylinder body through the bracket to the oil element, is filtered. The oil pressure is regulated, and lube oil is fed back to main gallery in the cylinder body.
The lube oil which flows in the main gallery goes to the crankshaft journal, lubricates the crank pin from the crankshaft journal, and a option of the oil is fed to the camshaft bearings.
Oil is sent from the gear case camshaft bearings through the holes in the cylinder body and cylinder head to the valve arm shaft to lubricate the valve arm and valves.
Oil is also sent through the intermediate gear bearing (oil) holes to lubricate the intermediate gear bearings and respective gears.
Lube oil for the fuel injection pump is sent by pipe from the main gallery to the fuel injection pump.

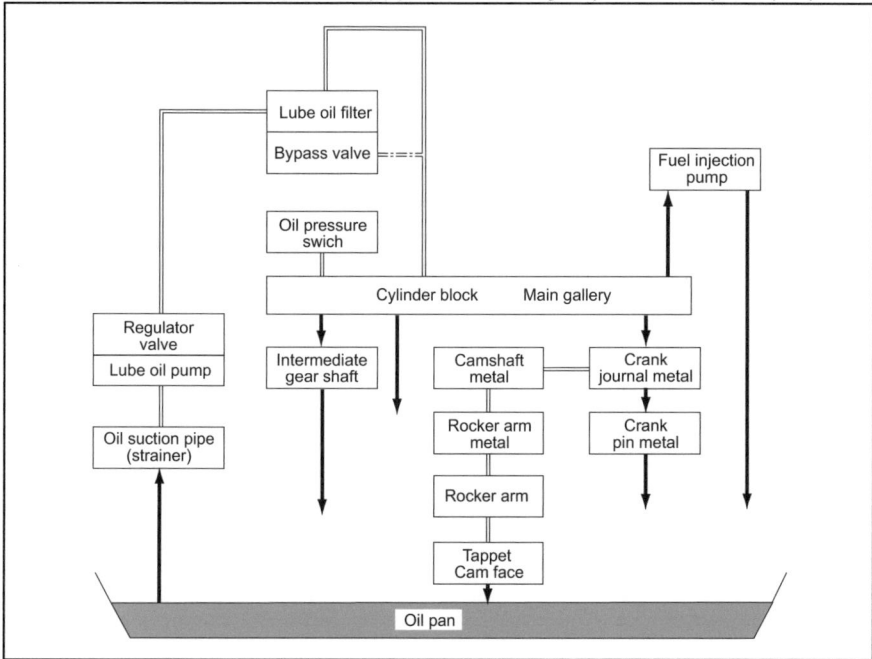

8.2 Lube oil pump

8.2.1 Lube oil pump construction

The trochoid type lube oil pump is mounted in the gear case cover, and the inner rotor is driven by the crankshaft pulley.

The lube oil flows from the intake filter mounted on the bottom of the cylinder body through the holes in the cylinder body and engine plate, and out from the holes in the engine plate and cylinder body to the discharge filter.

The lube oil pump is fitted with a control valve, which controls the discharge oil pressure at the specified pressure.

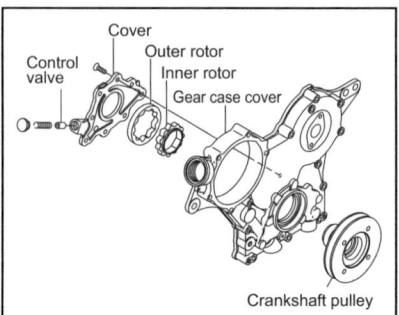

Cover
Outer rotor
Control valve
Inner rotor
Gear case cover
Crankshaft pulley

8.2.2 Specifications of lube oil pump

Lube oil pump specifications

Engine speed	3600 (min^{-1})	800 (min^{-1})
Pump speed	3477 (min^{-1})	772 (min^{-1})
Delivery quantity	$\geqq$19.0 (l/ min^{-1})	$\geqq$8.0 (l/ min^{-1})
Delivery pressure	0.43 MPa (4.4 kgf/cm^2)	$\geqq$0.049 MPa (0.5 kgf/cm^2)
Oil temp.	60±5 (°C)	←

8.2.3 Lube oil pump disassembly and reassembly

Disassembly

(1) Remove the crankshaft pulley.

(2) Remove the gear case cover.

(3) Remove the lube oil pump cover from the gear case cover. Do not disassemble the inner /outer rotors, and check that the pump rotates smoothly.

(4) Remove the pressure control valve from the lube oil pump cover.

Note :
Only wash the control valve. Disassembly is unnecessary unless any abnormality in operation is detected.

Reassembly

[NOTICE]
Always check if the pump rotates smoothly after installation on the gear case.
Running the engine when the pump rotation is heavy may cause the pump to be burnt.

(1) Apply lube oil to rotor (outer/inner) insertion part.

(2) Fasten the pump cover by the standard torque.

Tightening torque	5.9-7.9 N•m (0.6-0.8 kgf•m)

(3) When replacing the lube oil pump, replace the whole assy.

8. Lubrication system

8.2.4 Lube oil pump inspection

(1) Outside diameter clearance and side clearance of outer rotor

Insert a gap gauge between the outer rotor and the gear case cover, and measure outside diameter gap.
Put a ruler on the end face of the gear case cover, and insert a gap gauge between rotor, and measure a side gap.

Outside clearance   mm

Standard	Limit
0.12-0.21	0.30

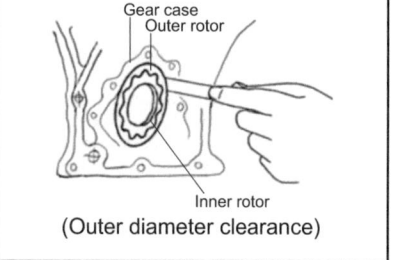
(Outer diameter clearance)

(2) Tip clearance between outer rotor and inner rotor

Insert a gap gage between an outer rotor and an inner rotor, and measure the tip clearance.

Tip clearance   mm

Standard	Limit
-	0.16

(3) Side clearance

When measuring a side clearance, put a right-angle gage to the pump body, insert a gap gage and measure the clearance.

Side clearance   mm

Standard	Limit
0.02-0.07	0.12

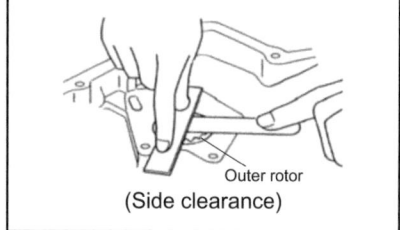
(Side clearance)

(4) Outside diameter clearance of inner rotor centering location part

Measure the outside diameter of inner rotor centering location part and the hole diameter of gear case cover. Calculate the clearance from that difference.

Inspection item	Standard	Limit
Gear case cover I.D.	46.13-46.18	-
Inner rotor O.D.	45.98-46.00	-
Rotor clearance	0.13-0.20	0.25

8.2.5 Pressure control valve construction

The pressure control valve controls the oil pressure from the time the lube oil leaves the filter.
When the pressure of lube oil entering the cylinder body main gallery exceeds the standard, the control valve piston opens the bypass hole and lube oil flows back into the oil pan.

Standard oil pressure	0.29-0.44 MPa (3.0-4.5 kgf/cm^2)

8.3 Lube oil filter

8.3.1 Lube oil filter construction

The lube oil filter is a full-flow paper element type mounted to the side of the cylinder block. The cartridge type filter is easy to remove.
To prevent seizure in the event of the filter clogging up, a bypass circuit is provided in the oil filter. When the difference of the pressure in front and behind the paper element reaches 0.08-0.12 MPa (0.8-1.2 kgf/cm^2), the bypass valve inside the filter opens and the lube oil is sent to each part automatically as an emergency measure, without passing through the filter.

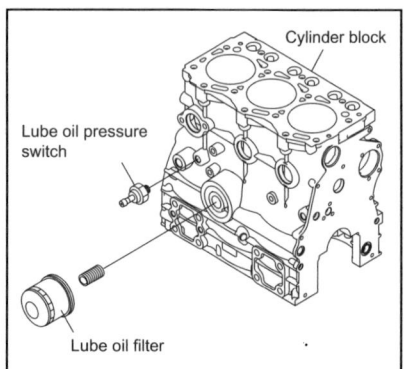

Type	Full flow, paper element
Filtration area	0.10 m^2
Discharge volume	30 l/min
Pressure loss	0.03-0.05 MPa (0.3-0.5 kgf/cm^2)
By-pass valve regulating pressure	0.08-0.12 MPa (0.8-1.2 kgf/cm^2)

8.3.2 Lube oil filter replacement

Refer to 2.2.2(2).

8. Lubrication system

8.4 Rotary waste oil pump (Optional)

A rotary waste oil pump to pump out waste oil is available as an option.

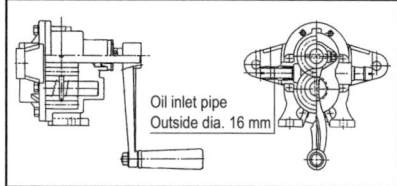

Oil inlet pipe
Outside dia. 16 mm

Rotary waste oil pump

Delivery capacity per one revolution	0.057 L
Delivery pressure	0.15 MPa (1.5 kg/cm^2) or below
Suction head	less than 1m
Part No.	124413-39100

9. Cooling water system

9.1 Cooling water system

The cooling water system is of the indirect seawater cooled, fresh water circulation type. The cylinders, cylinder heads and exhaust manifold are cooled with fresh water, and fresh water cooler (heat exchanger) use seawater.
Seawater pumped in from the sea by the seawater pump goes to the heat exchanger, where it cools the fresh water. Then it is sent to the mixing elbow and is discharged from the ship with the exhaust gas.
Fresh water is pumped by the fresh water pump to the cylinder jacket to cool the cylinders and the cylinder head. The fresh water pump body also serves as a discharge passageway (line) at the cylinder head outlet and is fitted with a thermostat.
The thermostat is closed when the fresh water pump temperature is low, immediately after the engine is started and during low load operation, etc. Then the fresh water flows to the fresh water pump inlet, and is circulated inside the engine without passing through the heat exchanger.
When the temperature of the fresh water rises, the thermostat opens, fresh water flows to the heat exchanger, and it is then cooled by the seawater in the tubes as it flows through the cooling pine. The temperature of it flows through the cooling pine. The temperature of the fresh water is thus kept within a constant range by the thermostat.

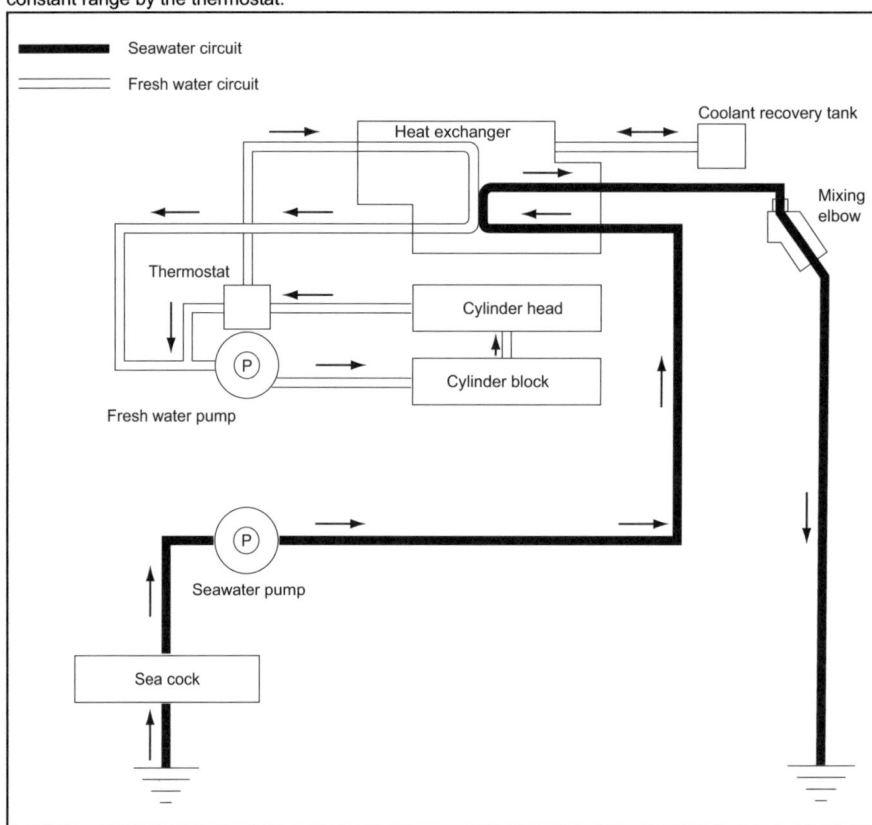

9. Cooling water system

Cooling water line

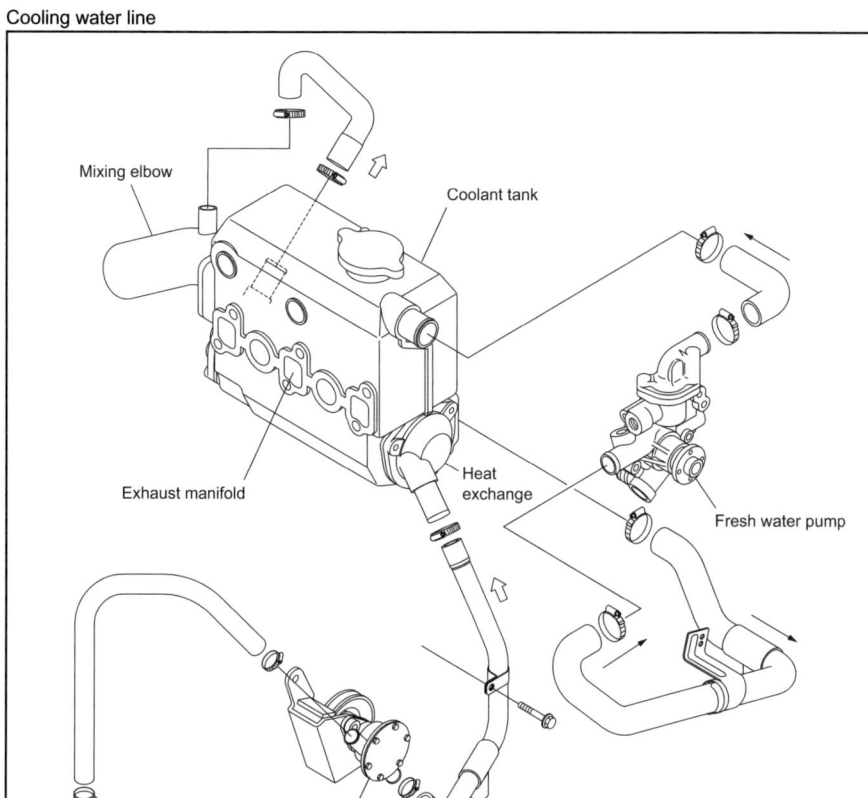

(Cooling water pipes)

9.2 Seawater pump

The seawater pump is driven by a V-belt.

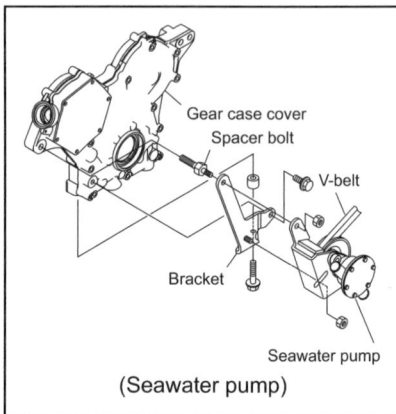

(Seawater pump)

9.2.1 Specifications of seawater pump

Performance

| Flow | Min. 1650 L/h at engine speed 3600 min^{-1} |

9.2.2 Seawater pump disassembly

Refer to 2.2.5.(5).

(1) Remove the rubber hose from the seawater pump outlet and then remove the seawater pump assembly from the gear case cover.

(2) Remove the seawater pump cover and take out the O-ring, and impeller.

(3) Remove the oil seal and the pump shaft if necessary.

9.2.3 Seawater pump Inspection

(Refer to 2.2.4(5).)

(1) Inspect the rubber impeller, checking for splitting around the outside, damage or cracks, and replace if necessary.

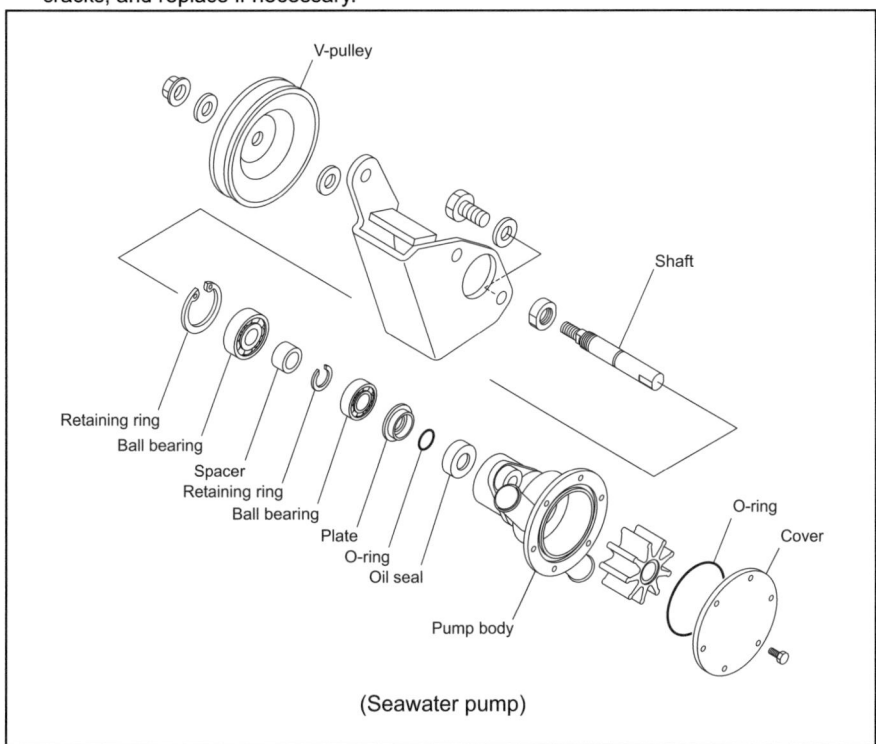

(Seawater pump)

(2) Inspect the oil seal and replace if it is damaged. Also replace the oil seal if there is considerable water leakage during operation.

Cooling water leakage	less than 3 cm^3/h

(3) Make sure the ball bearings rotate smoothly. Replace if there is excessive play.

9.2.4 Seawater pump reassembly

(1) When replacing the oil seal, coat with grease and insert.

(2) Mount the pump shaft, ball bearing and V-pulley to the pump unit and fit the bearing retaining ring.

NOTE:
   Coat the shaft with grease.

(3) Mount the impeller.

(4) Mount the O-ring and the cover.

NOTE:
   Replace the O-ring with new one.

9.3 Fresh water pump

9.3.1 Fresh water pump construction

The fresh water pump is of the centrifugal (volute) type, and circulates water from the fresh water tank to the cylinder block and the cylinder head.
The fresh water pump consists of the pump body, impeller, pump shaft, bearing unit and mechanical seal. The V-pulley on the end of the pump shaft is driven by a V-belt.
The bearing unit assembled in the pump shaft uses grease lubricated ball bearings and cannot be disassembled.

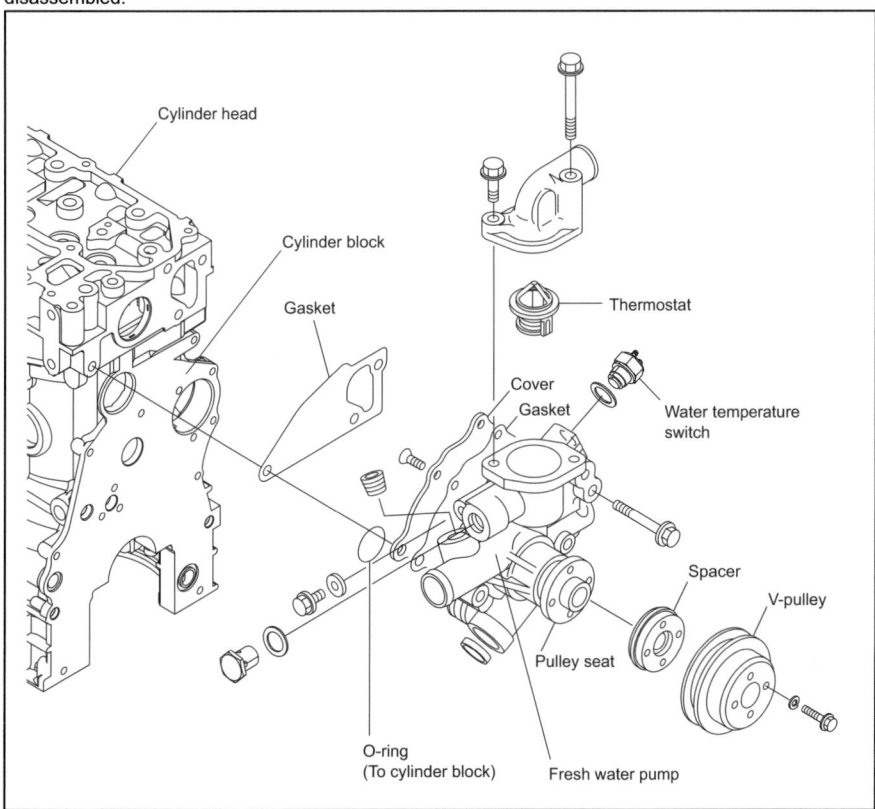

9.3.2 Specifications of fresh water pump

Pump shaft speed	3860 ± 30 min^{-1}
Delivery capacity	66.7 L/min or more
Total head	4 m Aq

9. Cooling water system

9.3.3 Fresh water pump disassembly

(1) Do not disassemble the fresh water pump. It is difficult to disassemble and, once disassembled, even more difficult to reassemble. Replace the pump as an assembly in the event of trouble. When the fresh water pump and the cover are disassembled, retighten to the specified torque.

Tightening torque	9.3-11.3 N•m (0.95-1.2 kgf•m)

(2) When removing the fresh water pump, replace the O-ring (inlet to cylinder).

9.3.4 WFresh water pump inspection

(1) Bearing unit inspection
Rotate the impeller smoothly. If the rotation is not smooth or abnormal noise is head due to excessive bearing play or contact with other parts, replace the pump as an assembly

(2) Impeller inspection
Check the impeller blade, and replace if damaged or corroded or if the impeller blade is worn due to contact with pump body.

(3) Check the holes in the cooling water and bypass lines, clean out any dirt or other foreign matter and repair as necessary.

(4) Replace the pump as an assembly if there is excessive water leakage due to mechanical seal or impeller seal wear or damage.

(5) Inspect the fresh water pump body, clean off scale and rust, and replace if corroded.

(6) Measure the outside clearance between the impeller and the pump body, by pushing the impeller all the way towards the body, and inserting a thickness gauge diagonally between the impeller and the body.
Measure the side clearance between the impeller and the plate by placing a straight-edge against the end of the pump body and inserting a thickness gauge between the impeller and the straight-edge.

Measuring outside clearance between impeller and pump body.

mm

	Standard	Limit
Outside clearance between impeller and body	0.3-1.1	1.5

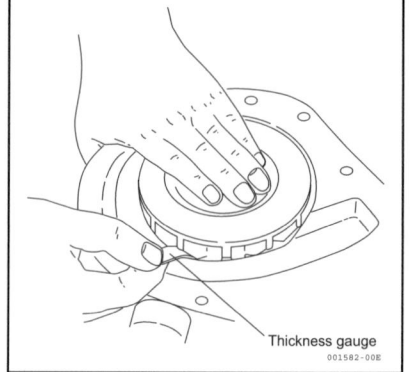

Thickness gauge

9. Cooling water system

Measuring side clearance between impeller and plate.

mm

	Standard	Limit
Side clearance between impeller and plate	0.5	-

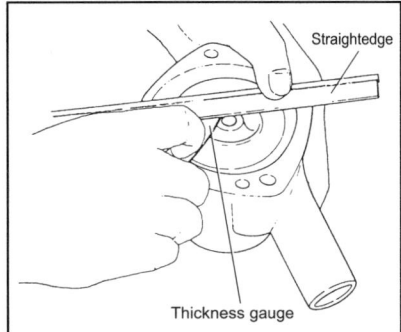

9.4 Heat exchanger

9.4.1 Heat exchanger construction

The heat exchanger cools the hot fresh water, that has cooled the inside of the engine, with seawater.
The cooler core consists of many small diameter tubes, baffle plates and tube cover.
The seawater flows through the maze formed by the baffle plates.
There is a reservoir above the cooler core, which serves as the fresh water tank.
There is an exhaust gas passageway in the reservoir, which forms a water cooled exhaust manifold.
The pressure cap (filler cap) on top of the heat exchanger has a pressure valve, which lets off steam through the overflow pipe when pressure in the fresh water system exceeds the specified value. It also takes in air from the overflow pipe when pressure in the fresh water system drops below the normal value.

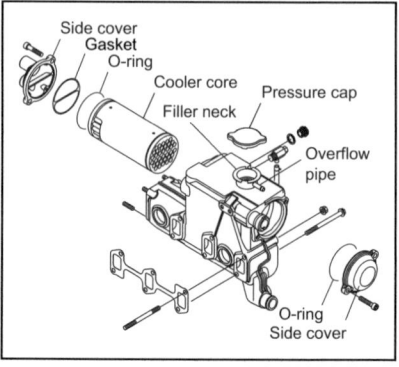

9.4.2 Disassembly and reassembly of the heat exchanger

(1) Remove the side covers on both sides of the cooler core and take out the cooler core and O-ring(s).

NOTE:
  Replace the O-ring(s) when you have removed the cooler core.

(2) Remove the pressure cap.

9.4.3 Heat exchanger inspection

Refer to 2.2.6(4).

9.5 Pressure cap and coolant recovery tank

9.5.1 Pressure cap construction

The pressure cap mounted on the filler neck incorporates a pressure control valve. The cap is mounted on the filler neck cam by placing it on the rocking tab and rotating. The top seal of the cap seals the top of the filler neck, and the pressure valve seals the lock seat.

9.5.2 Pressure cap pressure control

The pressure valve and the vacuum valve seal both the valve seats, when the pressure in the fresh water system is within the specified value of 88 kPa (0.9 kgf/cm^2). This seals the fresh water system.
When the pressure within the fresh water system exceeds the specified value, the pressure valve opens, and steam is discharged through the overflow pipe. When the fresh water is cooled and the pressure within the fresh water system drops below the normal value, atmospheric pressure opens the vacuum valve, and air is drawn in through the overflow pipe.

The coolant recovery tank (which sill be described later), keeps the water level from dropping due to discharge of steam when the pressure valve opens.

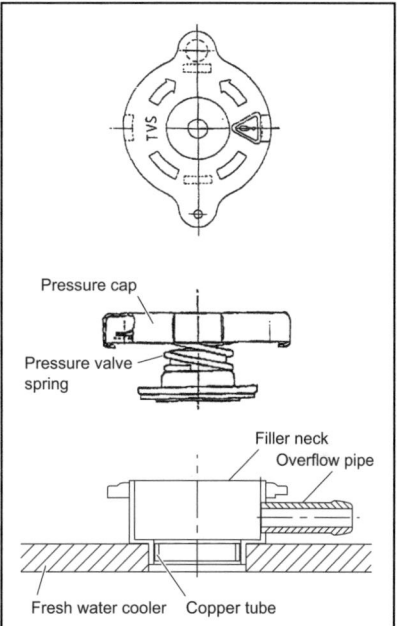

Action of pressure control valve

Pressure valve	Open at 82-109 kPa (0.84-1.11 kgf/cm^2, 0.82-1.09 bar)
Vacuum valve	Open at 8 kPa (0.08 kgf/cm^2) or below

9.5.3 Pressure cap inspection

Refer to 2.2.6(4).

9. Cooling water system

9.5.4 Replacing filler neck

1) Take out the copper pipe inside the filler neck with striking a circumference with a driver and so on. When the filler neck is removed, remove it with being careful not to damage the fresh water cooler, and scrap it.
2) Clean both new filler neck and the insertion part of fresh water cooler. Apply T7471 type activator or equivalent on both the surfaces and let it evaporate.
3) Apply Loctite 603 (improved 601) glue or equivalent on filler neck outside contour and press the filler neck into the fresh water cooler with the special tool.

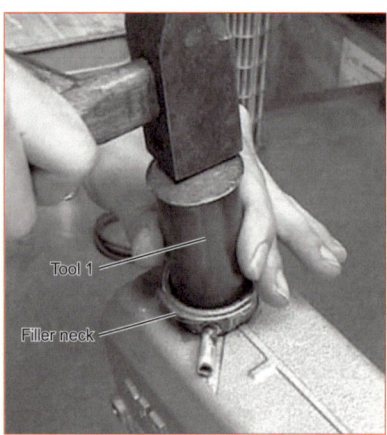

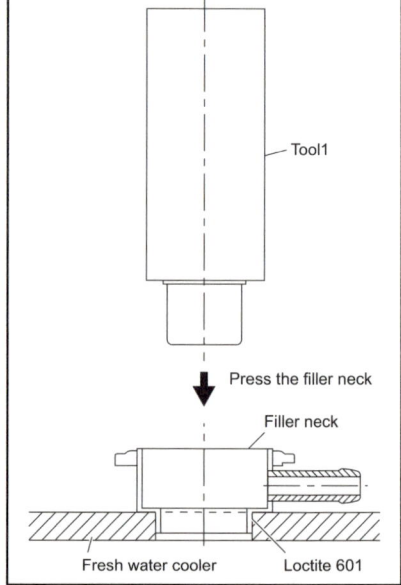

4) To fix the filler neck on the fresh water cooler, press the small copper tube inside the filler neck with the special tool.

Note:
The top of this tube should be under the sealing surface of the filler neck for the pressure cap.

5) Fit the pressure cap on the filler neck.

Filler neck Part No.	Copper tube Part No.
129673-44110	129673-44150

Refer to 4.2.2 for tool 1 and tool 2.

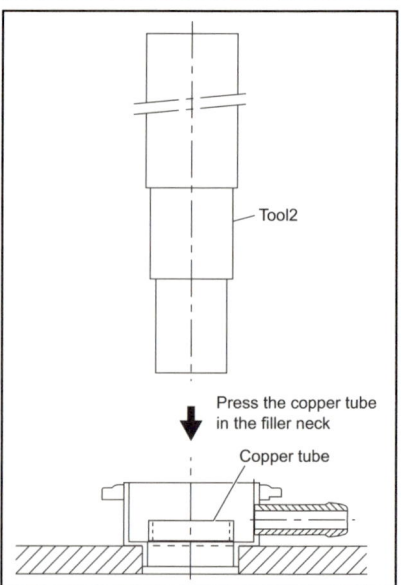

9. Cooling water system

9.5.5 Function of the coolant recovery tank

The pressure valve opens to discharge steam when the steam pressure in the fresh water tank exceeds 90 kPa (0.9 kgf/cm^2).
This consumes water. The coolant recovery tank maintains the water level by preventing this discharge of water.
The steam discharged into the coolant recovery tank condenses into watch, and the water level in the tank rises.
When the pressure in the fresh water system drops below the normal value, the water in the coolant recovery tank is sucked back into the fresh water tank to raise the water back to its original level.
The coolant recovery tank facilitates long hours of operation without water replacement and eliminates the possibility of burns when the steam is ejected from the filler neck because the pressure cap does not need to be removed.

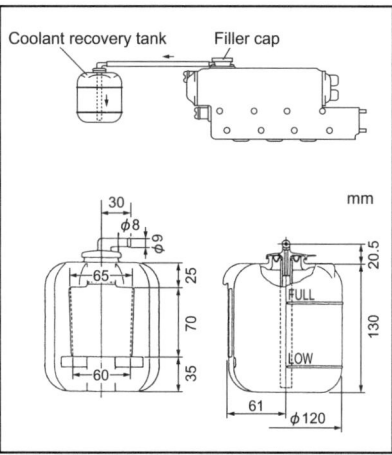

9.5.6 Specifications of coolant recovery tank

Capacity of coolant recovery tank	Overall capacity	1.3 L
	Full-scale position	0.8 L
	Low-scale position	0.2 L

9.5.7 Mounting the coolant recovery tank

(1) The coolant recovery tank is mounted at approximately the same height as the heat exchanger (fresh water tank).
(allowable difference in height : 300 mm (11.8110 in.) or less)

(2) The overflow pipe should be less than 1000 mm (39.3701 in) long, and mounted so that it does not sag or bend.

NOTE : Make sure that the overflow pipe of the coolant recovery tank is not submerged in bilge. If the overflow pipe is submerged in bilge, water in the bilge will be siphoned into the fresh water tank when the wafer is being cooled.

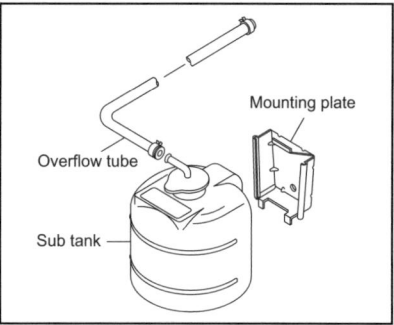

9.5.8 Precautions on usage of the coolant recovery tank

(1) Check the coolant recovery tank when the engine is cool and refill with fresh water as necessary to bring the water level between the low and full marks.

(2) Check the overflow pipe and replace if bent or cracked. Clean out the pipe if it is clogged up.

9. Cooling water system

9.6 Thermostat

9.6.1 Functioning of thermostat

The thermostat opens and closes a valve according to changes in the temperature of the fresh water inside the engine, controlling the volume of water flowing to the heat exchanger from the cylinder head, and in turn maintaining the temperature of the fresh water in the engine at a constant level.
The thermostat is bottom bypass type. It is located in a position connected with the cylinder head outlet line at the top of the top of fresh water pump unit.
When the fresh water temperature exceeds the above temperature, the thermostat opens, and a portion of the water is sent to the heat exchanger and cooled by seawater, the other portion going from the bypass line to the fresh water pump intake.
The bypass line is closed off as the thermostat valve opens and is completely closed when the fresh water temperature reaches 81.5°C (valve lifts 4 mm (0.1575 in)), sending all of the water to the heat exchanger.

9.6.2 Thermostat construction

The thermostat used in this engine is of the wax pellet type, with a solid wax pellet located in a small chamber. When the temperature of the cooling water rises, the wax melts and increases in volume. This expansion and construction is used to open and close the valve.

9.6.3 Characteristics of thermostat

Opening temperature	69.5-72.5°C
Full open temperature	85°C
Valve lift at full open	8 mm or more

9.6.4 Thermostat inspection

Remove the thermostat cover on top of the fresh water pump and take out the thermostat. Clean off scale and rust and inspect, and replace if the characteristics (performance) have changed, or if the spring is broken, deformed ar corroded.

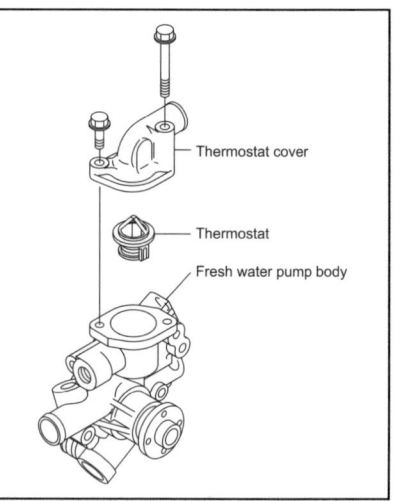

9.6.5 Testing the thermostat

Refer to 2.5 in chapter 2.

9. Cooling water system

9.7 Bilge pump and bilge strainer (Optional)

9.7.1 Introduction

(1) General Introduction

Name	Bilge pump
Time	10 minutes
Rotation direction	Right (viewed from the impeller side)
Weight	Pump 1.4 kg
Negative pressure detector	Diaphragm type
Temperature	-30°C~80°C

(2) Exterior

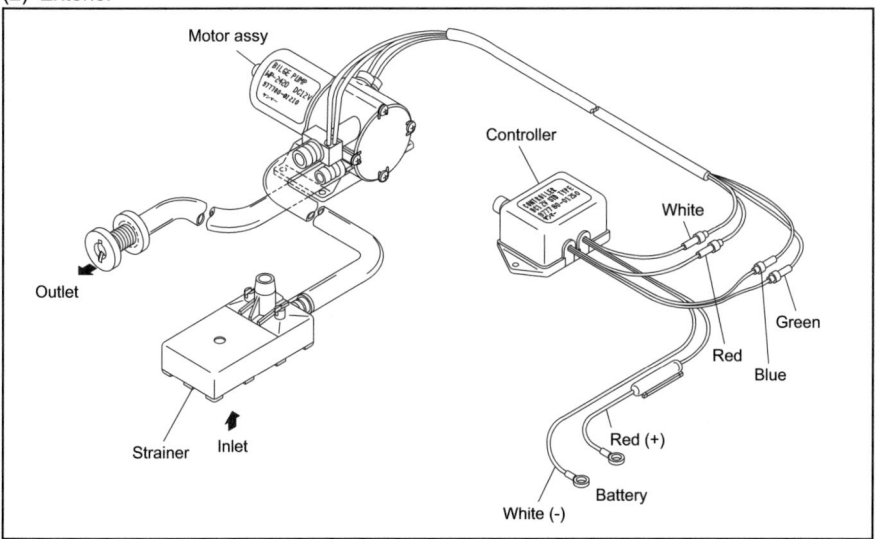

Pump dimensions

Length	225 mm
Yoke diameter	Ø61
Assembly hole diameter	Ø5.3
Assembly pitch	50 x 90 mm

9.7.2 Description

(1) Characteristics
   1) Discharge at lift : 0 m discharge capacity : 20 liters/min. or greater.
   2) Automatic feeding height : 1 m or greater
      (Limit for automatic feeding height: new pump with inside parts wet, approx. 2 m)
   3) Automatic feeding time : 2-5 seconds.
      (Limit for automatic feeding time: new pump with inside parts wet, approx. 1 second.)
   4) Automatic stopping : Air intake causes negative pressure triggering automatic stopping.

(2) Insulation
   1) Insulation resistance : 500V with a megatester when the difference between the continuity point and the body is 1MΩ or greater.
   2) Insulation proof stress : AC50 between the continuity point and the body, or 60hz 500V for 1 minute when impressed current leakage is 10 mA or lower.

(3) Durability
Rated voltage when there is 3% salt water 60L + engine oil 3%, and operation is at 1800 cycles and there are no difficulties.

(4) Vibration proof
Amplitude 0.51 mm (one side of the amplitude)
Vibration frequency 10-55 Hz
Sweep time 90 seconds.
Direction of vibration each direction 4 hours
No difficulties after test period

9.7.3 Cautions

(1) Attach at a position higher than the bilge water away from rain or other water, and 50-70 cm above the bottom of the boat.

(2) Never run the pump dry. Be sure that the strainer is inserted in the drain water before pushing the switch. If no water is being drawn up after a period of 10 seconds or more, prime the pump. (Do not run the pump for longer than 10 seconds when no water is being drawn up.)

(3) When the pump has not been used for a long period of time, the inside of the pump will be dry and drawing ability will be lowered. Before reusing, clean the inside of the pump or prime it to insure that it is wet, and check to be sure that the pump is then operating correctly.

(4) When charging the diesel engine oil, wait a period of 30 minutes or longer from the time of stopping (oil temperature 20-70°C). Refrain from operation when the oil temperature is below 15°C, or above 50°C.

(5) When the bilge inside the pump or hose freezes, completely melt the water with a steaming towel before beginning operation. When the temperature inside the pump is low, it will take a longer amount of time for the pump to drain off the bilge.

(6) The impeller replacement kit includes one impeller and 3 washers for adjusting the side gap. If after replacing the impeller the pump does not drain, place side gap adjustment washers underneath the bottom plate to adjust. Select the number of washers used in accordance with the following. (When the pump is draining, the electric current load is about 10A for 12V and 5A for 24V. When there are too many washers, the electric current value will be too great and will blow a fuse.)

9. Cooling water system

(7) The pump cannot be used to drain off rain water or large amounts of flood water. The pump can be run continuously for a period of 10 minutes. After this time it must shut off for a period of 2 hours before reusing.

(8) Do not use the pump for showering.
If the pump outlet is deformed for showering, the increase in water pressure will increase the load on the motor and cause motor seizure.

(9) Fix the strainer so that it will not turn upside down or on its side.

(10) When sludge has built up in the bilge to be drained, position the strainer about 20 mm above the sludge. When the pump is stopped, be sure there is no sludge remaining inside the pump.

(11) The specific gravity for the battery fluid is 1.25 or more.

9.7.4 Assembly procedure

When bilge is being used, assemble in accordance with the following.

(1) Assembling the bilge pump
- Select a dry place above the bilge water level.
- Select the location for the bilge pump taking into consideration the length of the switch cable (approx. 3 m) and its attachment point, and the position of the battery.
- Position at a 45° angle as shown in the illustration with the nozzle facing up, and 50-70 cm from the bottom of the boat.

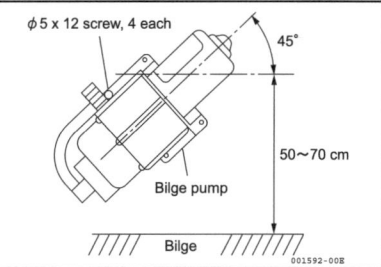

(2) Assembling the switch
- Attach in a place to insure easy operation away from rainwater.
- Connect the terminal to the battery.
(When the cord will not reach the battery, an extension of no greater than 3 m length suitable for AV3cm^2 can be attached.)

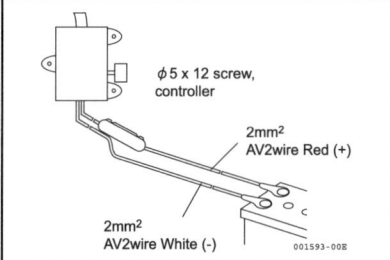

(3) Positioning the strainer
- Attach at the place where the greatest amount of water is collected when the boat is stopped.
- It is best to place the strainer as close to the bilge pump as possible. Cut the 3 m hose to a length of 1.2 m-1.8 m and attach allowing plenty of give.
- Check the strainer during a test operation before screwing firmly into place.
(When the strainer is screwed in, be especially careful not to damage the bottom of the boat.)
- The strainer contains a weight, and can be used with the weight in place.
- Always keep the strainer clean.

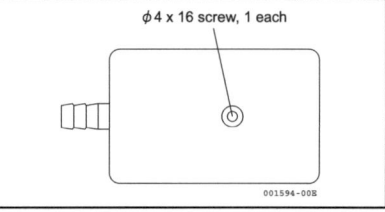

9. Cooling water system

(4) Attaching the delivery nozzle (outlet)
  - Make a fixing hole of Ø21 or less for attaching the nozzle. The hose attached at the nozzle should be 1.8 m or less and should reach without any strain, therefore care should be taken in deciding on the best position.
  - Fix the nozzle (outlet) in place and attach on the discharge side of the pump.

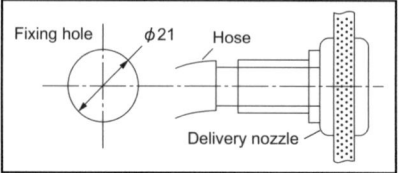

(5) Attaching the hose
  - Attach the hose from the strainer to the pump inlet.
  - Attach the delivery nozzle hose to the pump outlet.
  - Make the hose as short as possible and avoid sharp bends.

(6) Test operation
  - Collect water in the bottom of the boat, and check for any problems with the hose or wiring. After doing this, connect the battery.
  - Turn on the pump switch, and check to see that water is being taken in and discharged properly. The pump will stop automatically when there is no water left.
  - If the inside of the pump is dry, or if the water is not being drawn up initially after a period of 10 seconds, lift the strainer above the water surface and stop the pump. Prime the pump before starting it up again.

(7) Fixing the strainer
  - After the test operation, fix the strainer into place with screws.
    (Be careful not to damage the bottom of the boat with the screws.)

9. Cooling water system

9.7.5 Cautions for assembling

Observe the following cautions for handling.
- Do not use gasoline or solvents.
  1) gasoline 2) ester 3) benzol 4) battery fluid 5) liquids at 70°C or greater or engine oil
- Never run when there is no water in the bilge.
  Check to be sure that the strainer is in the water before turning on the switch.
- Keep the cord terminal away from the water. Water inside the motor or switch may lead to damage. When the insulation around the cord is damaged, water can seep in to the wires; thus, care should be taken not to scratch or nick the cord.
- When the pump has not been used for a long period of time, the inside of the pump will be dry and it may not operate properly at first. If after 10 seconds the pump is not working, turn off the switch and prime the pump before trying again.
  (Never run the pump dry for period of greater than 10 seconds.)

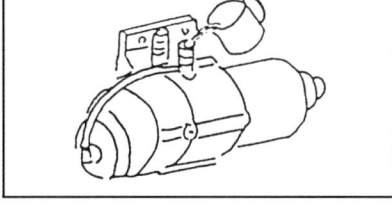

- Replace the diesel engine oil only after the engine has been stopped for a period of 30 minutes (oil temp. 20-70°C). Whenever possible refrain from operation when the oil temperature is below 15°C or above 50°C.
- Bilge water left in the hose or inside the pump can freeze, and care should be taken to see that any excess bilge is completely discharged. If bilge water should freeze, and care should be taken to see that any excess bilge is completely discharged. If bilge water should freeze inside the hose or pump, it should be completely melted before starting up the pump. When the temperature inside the pump is low, it will take a longer time for the pump to operate. (0°C, 5-10 seconds.)
- Keep the pump in a dry place away from rain or other water.
- Use the regulation hose; do no use thin vinyl hose or hose which is not heat-resistant.
- The pump cannot be used to drain off rainwater or large quantities of flood water. This pump can be operated continuously for a period of 10 minutes.
- Do not use the pump for showering.
  If the pump outlet is deformed for showering, the increase in water pressure will increase the load on the motor and cause motor seizure.
- When sludge has built up in the bilge to be drained, position the strainer about 20 cm above the sludge. When the pump is stopped, be sure there is sludge remaining inside the pump housing.
- The specific gravity for the battery fluid is 1.25.
- Refer to your local dealer for impeller replacement.
  The local dealer will perform the following.
  The impeller replacement kit includes one impeller and 3 films for adjusting the side gap. If after replacing the impeller the pump does not drain, place side gap adjustment washers underneath the bottom plate to adjust. Select the number of films used in accordance with the following. (When the pump is draining, the electric current load is about 10A for 12V and 5A for 24V. The pump operates efficiently at these electric current loads.)

Steps for replacement
1) Remove the impeller plate by taking out the M4 screws (4) and opening the top of the diaphragm switch.
   (Screw lock has been applied to the screw, and a dryer should be used to heat the screw before removing it.)
2) Clean the inside of the pump.
3) Grease the plate, impeller, and film for side gap adjustment, and then reassemble the pump by inserting first the film plate and then the impeller.

9. Cooling water system

9.7.6 Troubleshooting

Refer to the following countermeasures for difficulties that arise.

Problem	Cause	Countermeasure
1. Pump does not turn	Faulty wiring	Check the wiring between the motor and battery.
	Faulty battery	Check to see if the specific gravity of the battery fluid is greater than 1.25. Recharge or replace the battery.
	Faulty starter switch	Consult your local dealer.
	Faulty pump	Consult your local dealer.
2. Pump turns but does not draw up water.	Draws up air.	Check hose connections. Retighten pump screws.
	Low voltage in battery.	Check to see if the specific gravity of the battery fluid is greater than 1.25. Recharge or replace the battery.
	The distance between the pump and the surface of the water is too great.	Lower the pump. (Position the pump so that it is closer to the surface of the water.)
	The pump is too high.	Lower the pump. (Position the pump so that it is 50-70 cm above the bottom of the boat.)
	Pump intake is weak.	If intake is still faulty after priming, consult your local dealer.
3. Pump turns, but the amount of discharge is low.	Clogged strainer	Clean strainer.
	Hose is broken or damaged.	Check for damage and repair. If incorrect hose has been used, replace with the regulation type of hose.
4. Water leakage from pump	Water leakage from packing	Retighten pump screws.
	Faulty pump seal	Consult your local dealer.
5. Pump draws up bilge, but motor stops when hand is removed from starter switch.	Faulty diaphragm switch	Check for loose wiring in diaphragm switch and correct.
	Damaged diaphragm switch	Consult your local dealer.
6. Motor does not stop, when there is no bilge water left	Clogged strainer or hose	Clean strainer or hose.
	Damaged diaphragm switch	Check for continuity of diaphragm switch terminal. Consult your local dealer if there is continuity.

10. Reduction and reversing gear

Marine gear KM2P-1 is applied to the 3YM30, 3YM20 and 2YM15 series engines.

Refer to chapter 10 in the service manual of the GM series engines for inspection, disassembly and reassembly.

10.1 Specifications

Model			KM2P-1		
For engine models			3YM30, 3YM20, 2YM15		
Clutch			Constant mesh gear with servo cone clutch (wet type)		
Reduction ratio	Forward		2.21	2.62	3.22
	Reverse		3.06	3.06	3.06
Propeller shaft speed (at continuous power, Forward) $\min^{-1}$			1580	1332	1083
Direction of rotation	Input shaft		Counter-clockwise, viewed from stern		
	Output shaft	Forward	Clockwise, viewed from stern		
		Reverse	Counter-clockwise, viewed from stern		
Remote control	Control head		Single lever control		
	Cable		Morse. 33-C (cable travel 76.2 mm)		
	Clamp		YANMAR made. standard accessory		
	Cable connector		YANMAR made. standard accessory		
Output shaft coupling	Outer diameter		100 mm		
	Pitch circle diameter		78 mm		
	Connecting bolt holes		4 - 10.5 mm		
Position of shift lever, viewed from stern			Left side		
Lubricating oil			API CC class, SAE #10W30		
Lubricating oil capacity			0.3 liters		
Dry mass			9.8 kg		

11. Remote control (Optional)

11.1 Remote control system

11.1.1 Construction of remote control system

The remote control permits one handed control of the engine speed changing from forward to reverse, and stopping.
Fittings which allow for easy connection of the remote control cables with the fuel injection pump and transmission are provided with the remote control set.
The use of Morse remote control cables, clamps and a remote control head, are also provided for the device to stop the engine is electric and will be explained under the section on electrical equipment.

11.1.2 Remote control device components

	Morse description
Remote control head	Morse MT3 top mounting single lever Morse MN side mounting single lever
Remote control cable	Morse 33C x 4m Morse 33C x 7m
Engine stop cable	Yanmar 4m Yanmar 7m

(1) Remote control handle

The model MT-3 remote control has been designed so that operation of the clutch (shift) and governor (throttle) can be effected with one lever.
Two cables are required for the MT-2 single, one for the clutch and the other for the governor.
When warming up the engine, to freely control the governor separately from the clutch put the lever in-neutral, the central position and pull the knob in the center of the control lever. When the lever is returned to the neutral position, the knob automatically returns to its original position, and the clutch is free. The governor can then be freely operated.

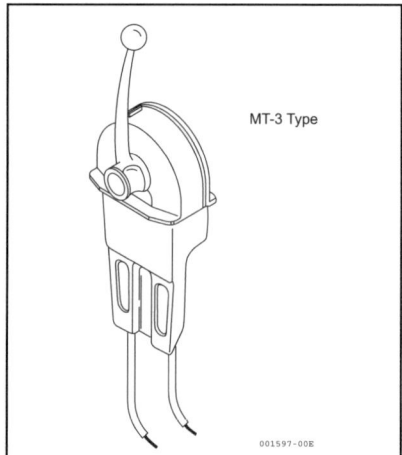

MT-3 Type

The MN type controller has been designed so that operation of the clutch and throttle can be effected with one lever. When the button next to the control lever is pulled out with the lever in the central position, it holds the clutch in the neutral position so that the throttle can be opened all the way and warm up the engine. When the engine is warmed up, return the handle to the central position and push the button back in. Control of the clutch and throttle is thus effected with one handle.

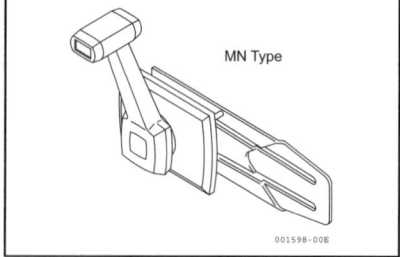

MN Type

11. Remote control (Optional)

(2) Remote control cable
Use only Super Responsive Morse Control Cables. These are designed specifically for use with Morse control heads. This engineered system of Worse cables, control head and engine connection kits ensures dependable, smooth operation with an absolute minimum of backlash.

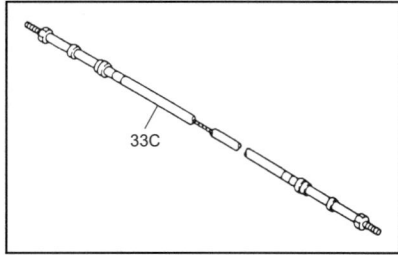

33C

(3) Engine stop cable

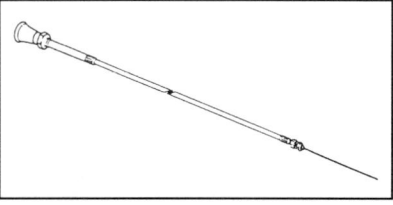

11. Remote control (Optional)

11.2 Remote control installation

(1) Speed control

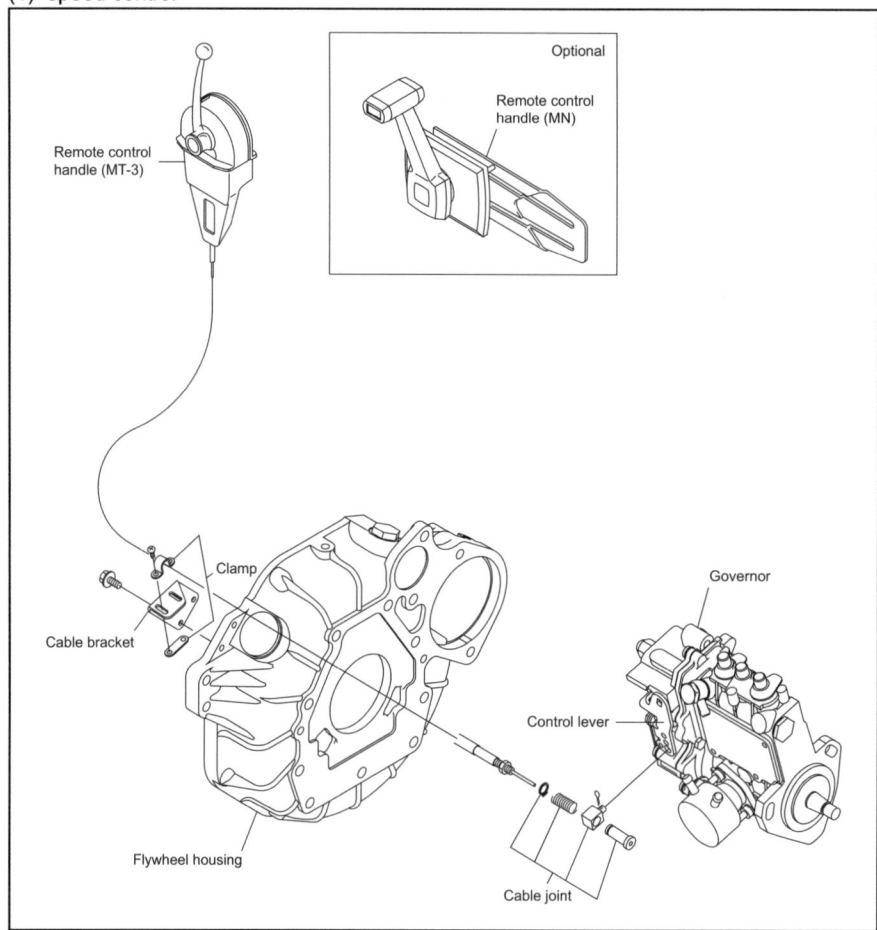

11. Remote control (Optional)

(2) Clutch control

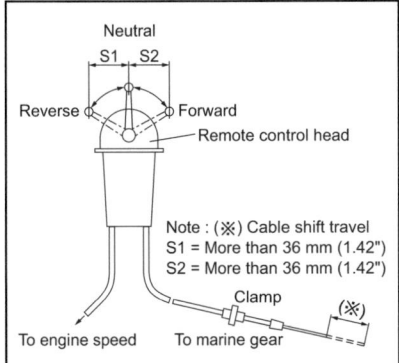

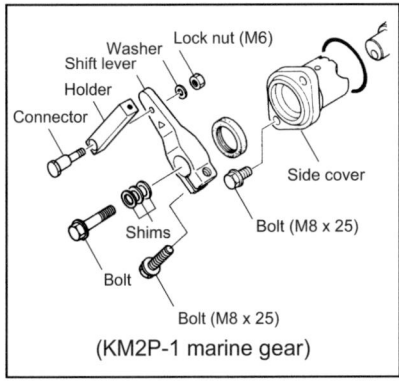

(KM2P-1 marine gear)

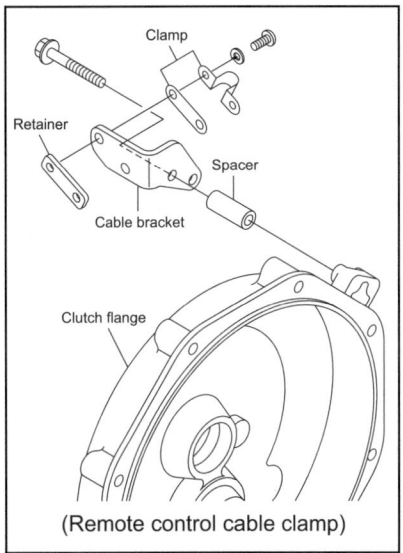

(Remote control cable clamp)

(3) Engine stop

Usually, when an engine is stopped, the stop button of an instrument panel is pushed, and the engine is stopped.
Moreover, an emergency stop button after the stop solenoid is pushed, and the engine is stopped.

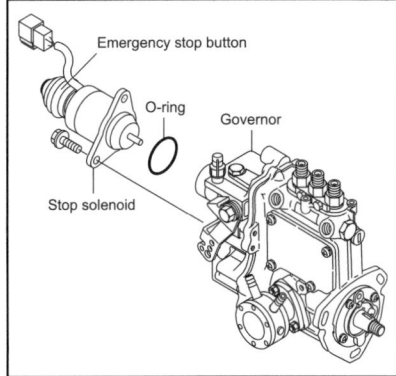

11.3 Remote control inspection

(1) When the control lever movement does not coincide with operation of the engine, check the cable end stop nut to see whether or not it is loose, and readjust/retighten when necessary.

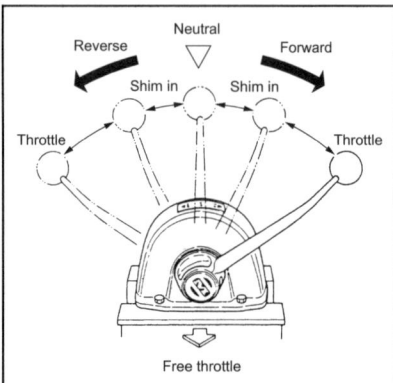

(2) Too many bends (turns) in the cable or bends at too extreme angle will make it difficult to turn the handle. Reroute the cable to reduce the number of bends and enlarge the bending radius as much as possible (to 200 mm or more).

(3) Check for loose cable bracket/clamp bolts or nuts and retighten as necessary.

(4) Check cable connection screw heads, cable sleeves and other metal parts for rust or corrosion. Clean off minor rust and wax or grease the parts. Replace if the parts are heavily rusted or corroded.

11.4 Remote control adjustment

(1) Shift lever adjustment
Move the lever several times-the movement of the clutch lever on the engine from forward, neutral and reverse must coincide with the forward, neutral and reverse on the control lever. If they do not coincide, adjust the fittings as necessary (first engine side, then controller side).

(2) Throttle lever adjustment
Move the control lever all the way to full throttle several times, and then return. The throttle lever on the engine must lightly push against the idle switch when it is returned. If it is property adjusted, the knob can be easily pulled out when the lever is in the neutral position, and will automatically return when the control lever is brought back to the neutral position. If the control lever presses too hard against the knob, it may not return automatically, in which case the cable end must be adjusted as explained for the clutch. The knob cannot be pulled out when the lever is not in the neutral (central) position.

12. Electrical system

12.1 Electrical system

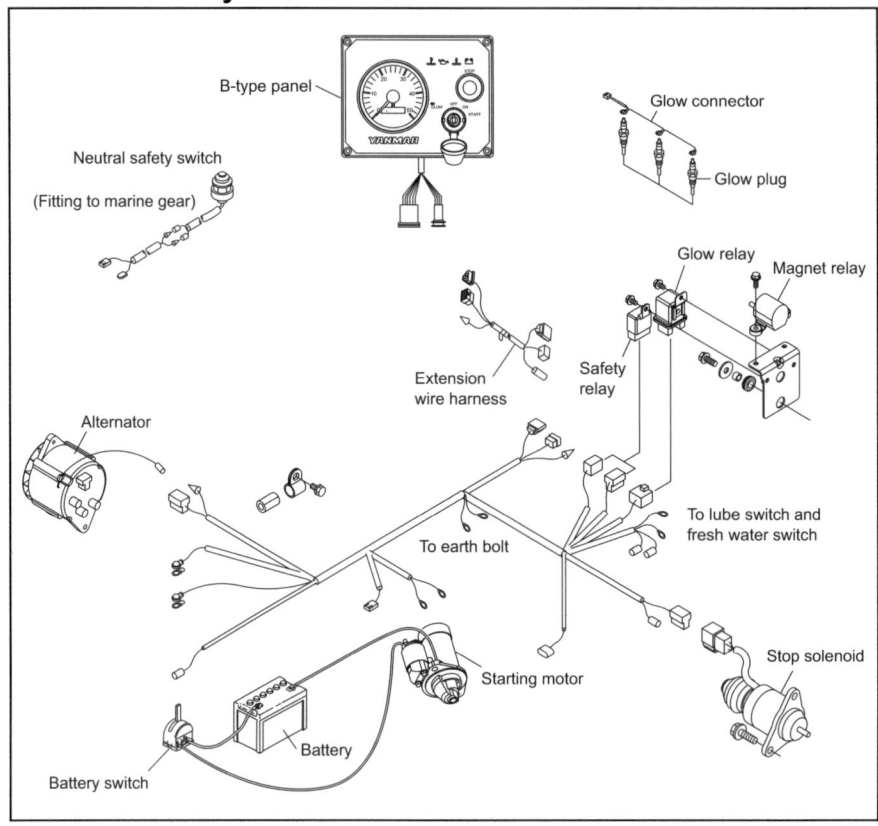

12.1.1 Wiring diagram

For B-type instrument panel

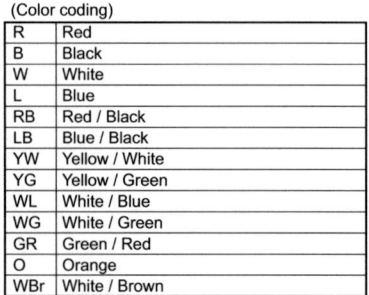

(Color coding)	
R	Red
B	Black
W	White
L	Blue
RB	Red / Black
LB	Blue / Black
YW	Yellow / White
YG	Yellow / Green
WL	White / Blue
WG	White / Green
GR	Green / Red
O	Orange
WBr	White / Brown

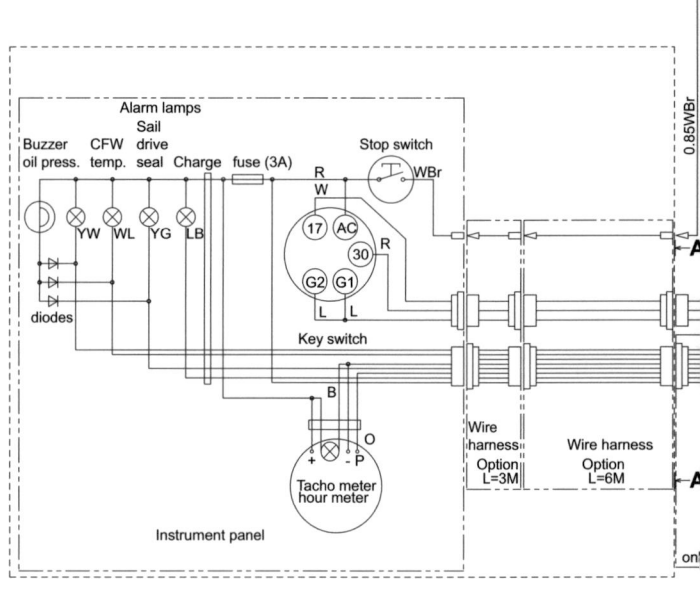

12. Electrical System

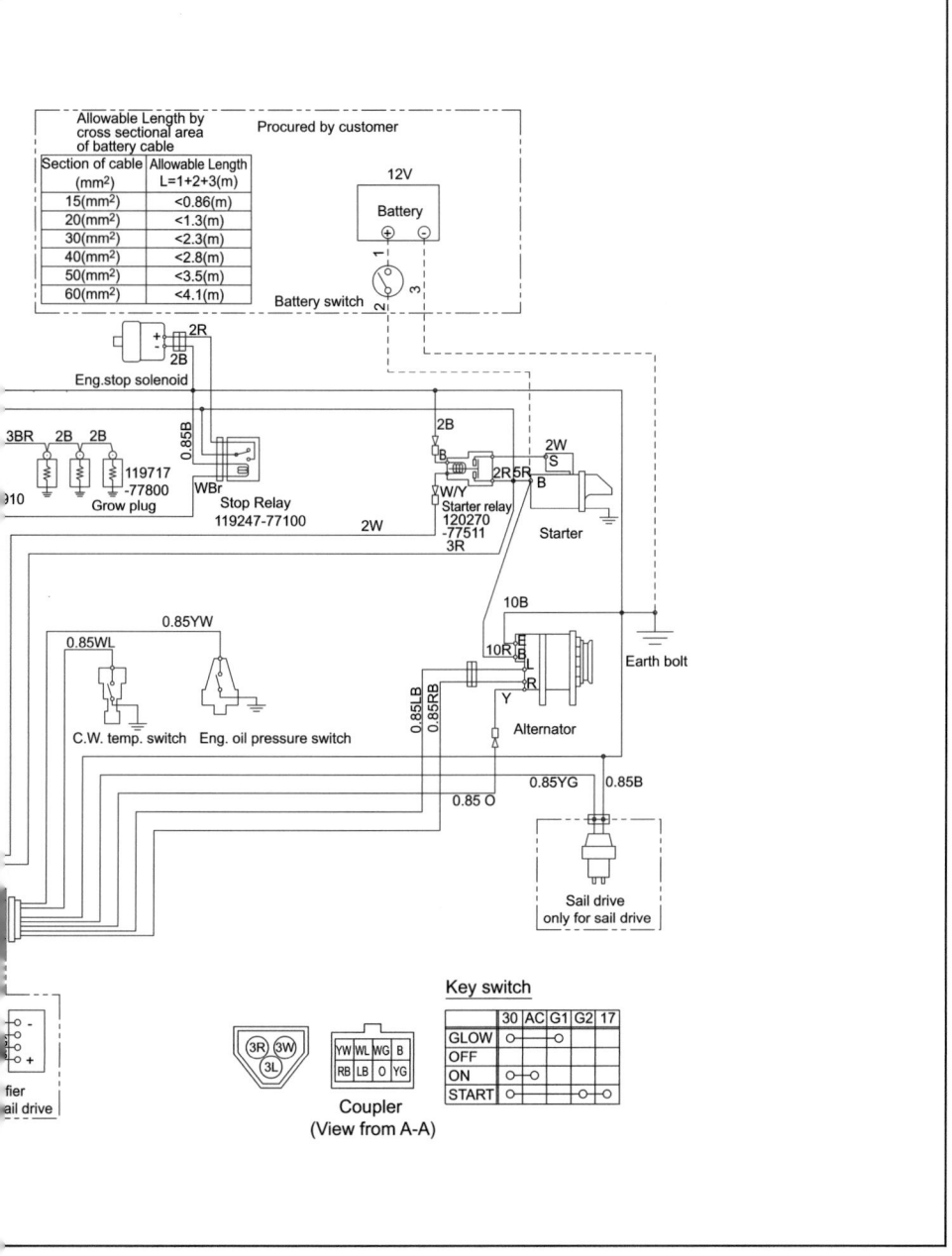

12. Electrical System

12.2 Battery

The new type instrument panels are applied for 3YM30 series engines. The features are compactness, waterproof and independence from pulse by ring gear teeth number.
The engine speed, indicated with the instrument panel is activated by the pulse from flywheel ring gear.
The engine speed with new panel is activated by alternator B terminal pulse.

(1) Battery capacity

Battery capacity (5 hours rating)	12V-64 AH or more (type 95D31L equivalent)

(2) Battery cable

Wiring must be performed with the specified electric wire. Thick, short wiring should be used to connect the battery to the starter. Using wire other than that specified may cause troubles.

The overall length of the wire between the battery (+) terminal and the starter (B) terminal, and between the battery (-) terminal and the starter (E) terminal, should be determined according to the following table.

Voltage system	Allowable wiring voltage drop	Conductor cross section area	Allowable overall length
12 V	0.2 V or less/100A	20 mm^2	Up to 2.5 m
		40 mm^2	Up to 5 m

Note :
 Excessive resistance in the key switch circuit (between the battery and start [S] terminals) can cause improper pinion engagement. To prevent this, follow the wiring diagram carefully.

12. Electrical System

12.3 Starting motor

A starting motor turns the ring gear installed on a engine flywheel by the pinion while overcoming resistance such as the compression pressure and the friction loss of the engine and makes the engine start.

12.3.1 Specifications

YANMAR Part No.	129608-77010
HITACHI Model No.	S114-817A
Nominal power (kW)	1.4
Nominal voltage (V)	12
Rating (sec)	30
Direction of rotation (Looking from the pinion side)	Clockwise
Number of pinion teeth	11
Weight (kg)	3.0
No load	Terminal voltage (V)	11
	Electric current (A)	90 (MAX)
	Revolutions (min^{-1})	2,700 (MIN)
Load	Terminal voltage (V)	8.4
	Electric current (A)	250
	Torque (N·m)	8.3 (MIN)
	Revolutions (min^{-1})	1,000 (MIN)

12.3.2 Characteristics

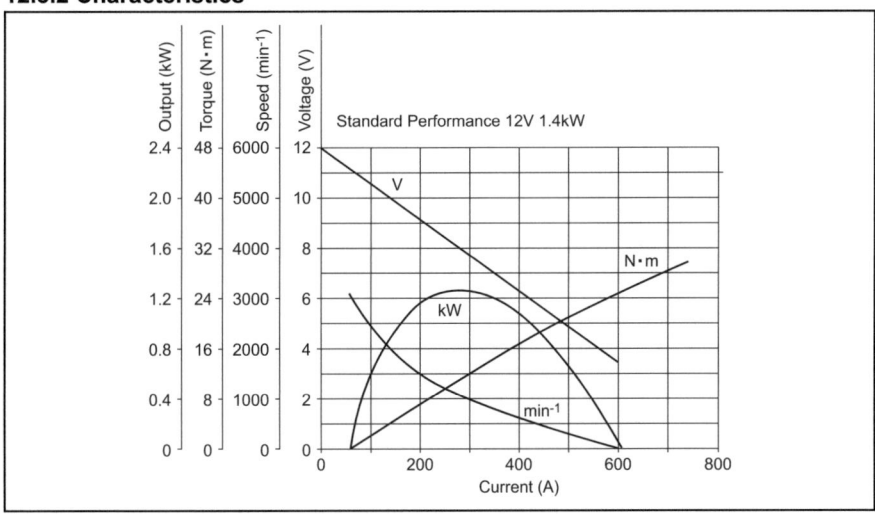

12.3.3 Structure

(1) Disassembly drawing

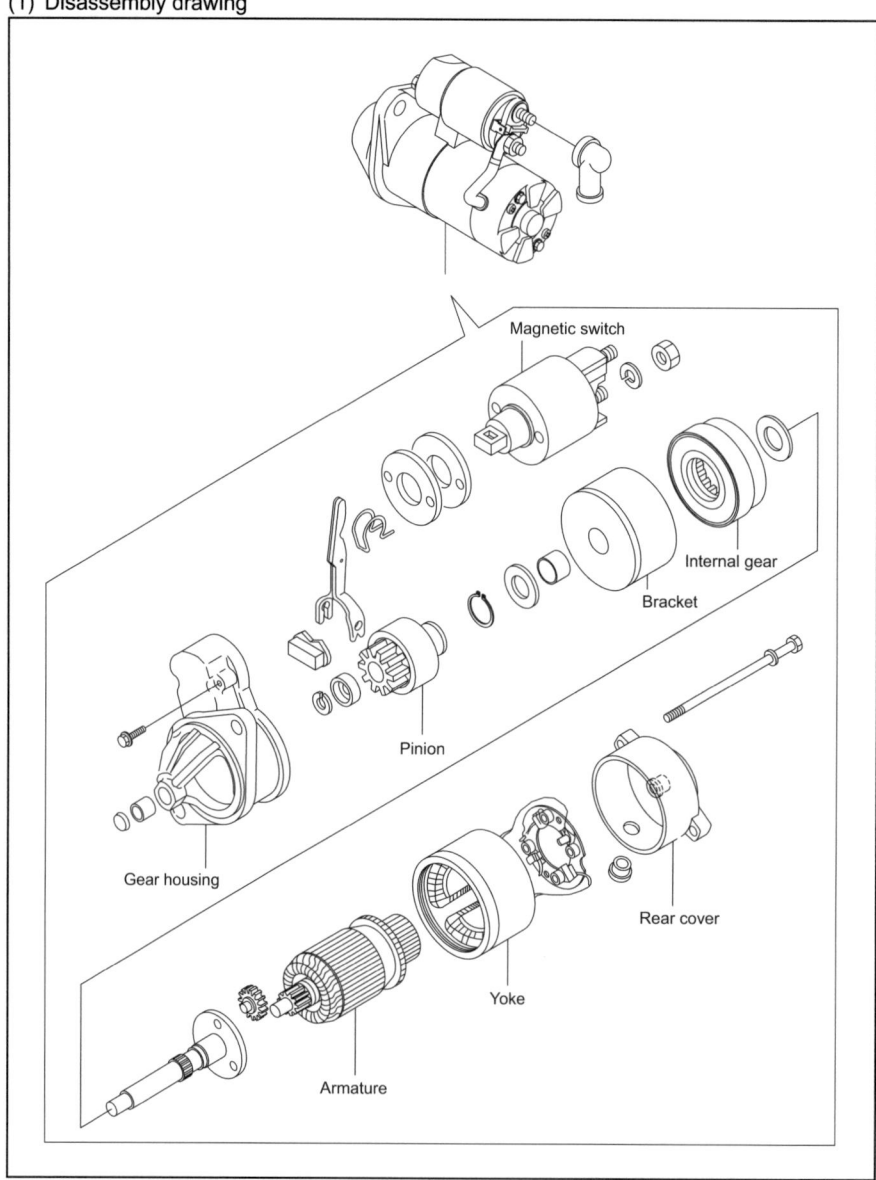

12. Electrical System

(2) Structure

When the starting switch is turned on, a magnet switch takes a voltage, and a pinion projects. The pinion engages with the ring gear of a engine, and the engine is started.

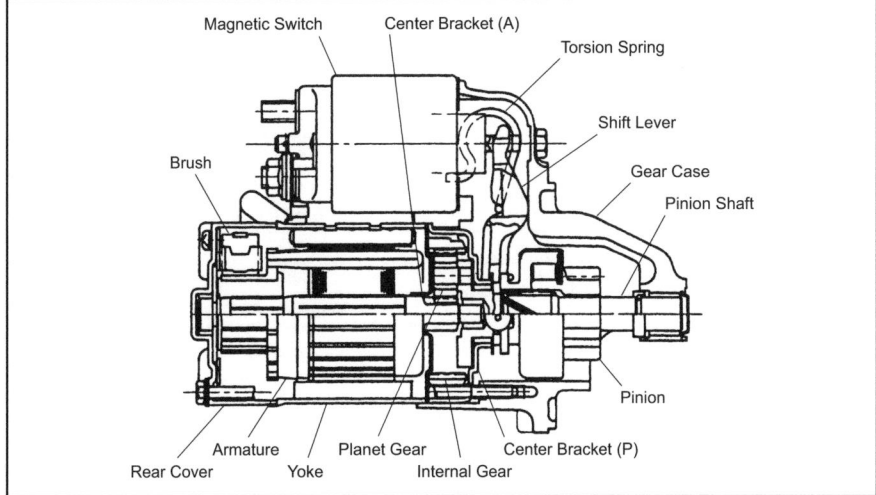

12.3.4 Wiring diameter of a starting motor

1) When a starting switch is turned on, a magnet switch is charged, and a moving core is absorbed, and a pinion clutch is moved forward through a lever, and the pinion engages with a ring gear.

2) When the pinion engages the ring gear, because a main contact point is closed and the main electric current flows and a pull coil is short-circuited by the main contact point and it stops being charged with electricity, the pinion is kept at the position by a holding coil during the start.

3) When the starting switch is turned off, the main contact point becomes open, and the pinion clutch is returned to the stop position by a return spring.

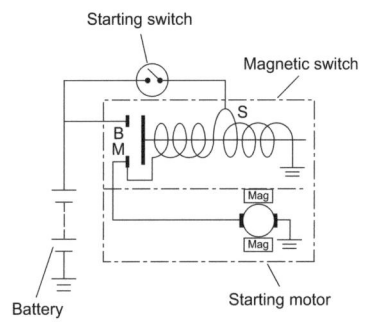

12.4 Alternator standard, 12V/60A

The alternator serves to keep the battery constantly charged. It is installed on the cylinder block by a bracket, and is driven from the V-pulley at the end of the crankshaft by a V belt.
The type of alternator used in this engine is ideal for high speed engines with a wide range of engine speeds. It contains diodes that convert AC to DC, and an IC regulator that keeps the generated voltage constant even when the engine speed changes.

12.4.1 Specifications

Yanmar code	128271-77200
Model of alternator	LR160-741 (HITACHI)
Model of IC regulator	SA-A (HITACHI)
Battery voltage	12V
Nominal output	12V/60A
Earth polarity	Negative earth
Direction of rotation (viewed from pulley end)	Clockwise
Weight	4.2 kg
Rated speed	5000 min^{-1}
Operating speed	1,050-18,000 min^{-1}
Speed for 13.5V at 20°C	1050 min^{-1} or less
Output current for 13.5V	56A or more/ 5000 min^{-1}
Regulated voltage	14.4±0.3V (at 20°C, voltage gradient, -0.01V/°C)

12. Electrical System

12.4.2 Structure

(1) Disassembly drawing

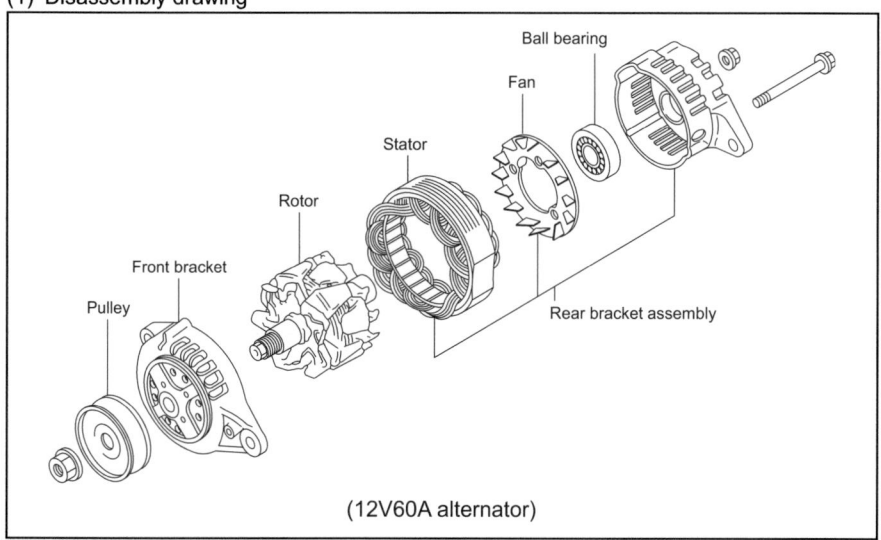

(12V60A alternator)

(2) Structure

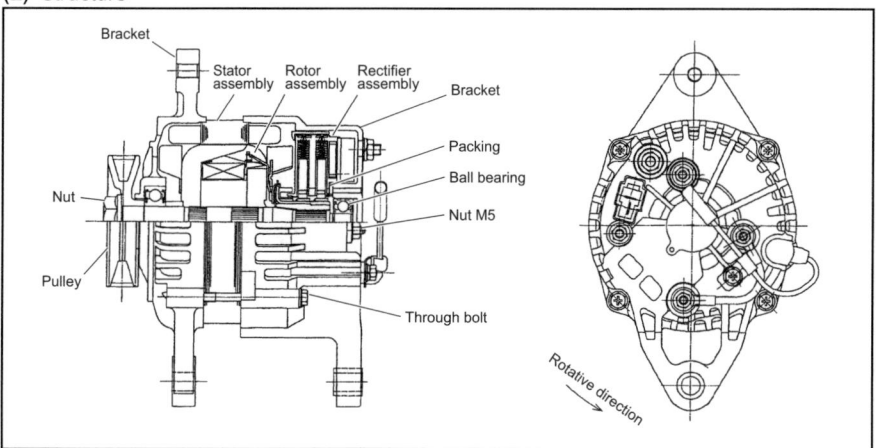

12.4.3 Wiring diagram

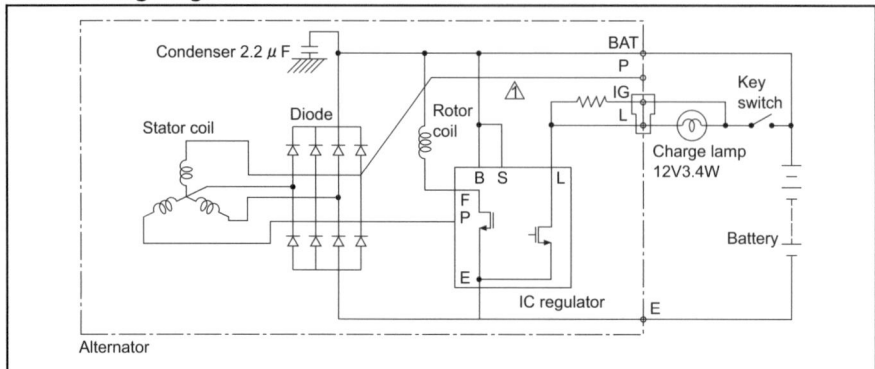

[NOTICE]
1) Don't do mis-connecting and short-circuit of each terminal.
2) Don't remove a battery terminal and a B terminal when rotating.
3) Shut out a battery switch during the alternator stop.

12.4.4 Standard output characteristics

The standard output characteristics of this alternator are shown as the below figure.

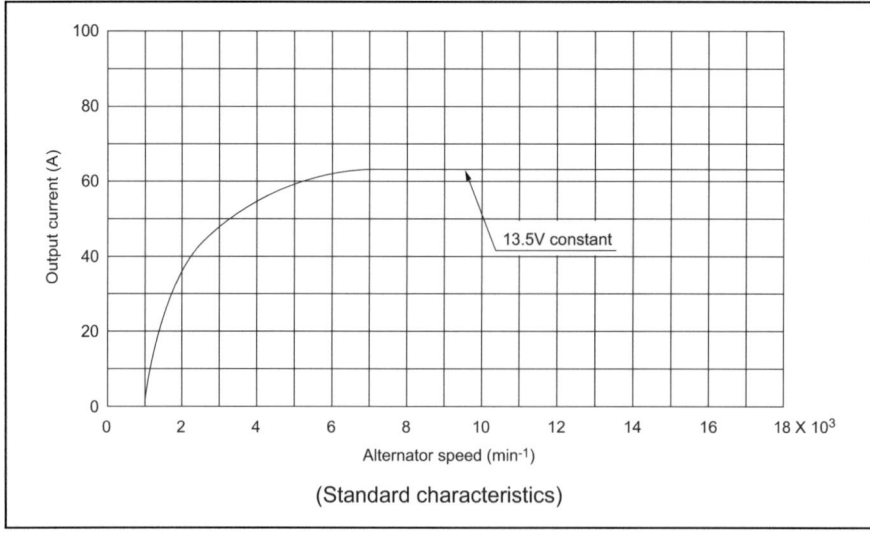

(Standard characteristics)

12.4.5 Inspection

(1) V belt inspection
   1) Inspect the matter whether there are not crack, stickiness and wear on the belt visually. Check that a belt doesn't touch the bottom part of the pulley groove. If necessary, replace the V belt set.
   2) V belt tension :
      (Refer to 2.2.2.(4) in Chapter 2.)

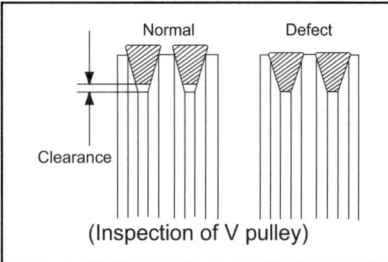
(Inspection of V pulley)

(2) Visual check of wiring and check of unusual sound
   1) Confirm whether wiring is right or there is no looseness of the terminal part.
   2) Confirm that there is no unusual sound from the alternator during the engine operation.

(3) Inspection of charge lamp circuit
   1) Move a start switch to the position of on. Confirm lighting of the charge lamp.
   2) Start an engine, and confirm the lights-out of the lamp. Repair a charge lamp circuit when a lamp doesn't work.

12.5 Alternator 12V/80A (Optional)

12.5.1 Specifications

Yanmar code	119573-77201
Model of alternator	LR180-03C (HITACHI)
Model of IC regulator	TR1Z-63 (HITACHI)
Battery voltage	12V
Nominal output	12V/80A
Earth polarity	Negative earth
Direction of rotation (viewed from pulley end)	Clockwise
Weight	5.4 kg
Rated speed	5000 min^{-1}
Operating speed	1,200-9,000 min^{-1}
Speed for 13.5V at 20°C	1,200 min^{-1} or less
Output current for 13.5V	75A or more/ 5000 min^{-1}
Regulated voltage	14.5 ± 0.3V (at 20°C, voltage gradient, -0.01V/°C)

12. Electrical System

12.5.2 Structure

Disassembly drawing and Structure

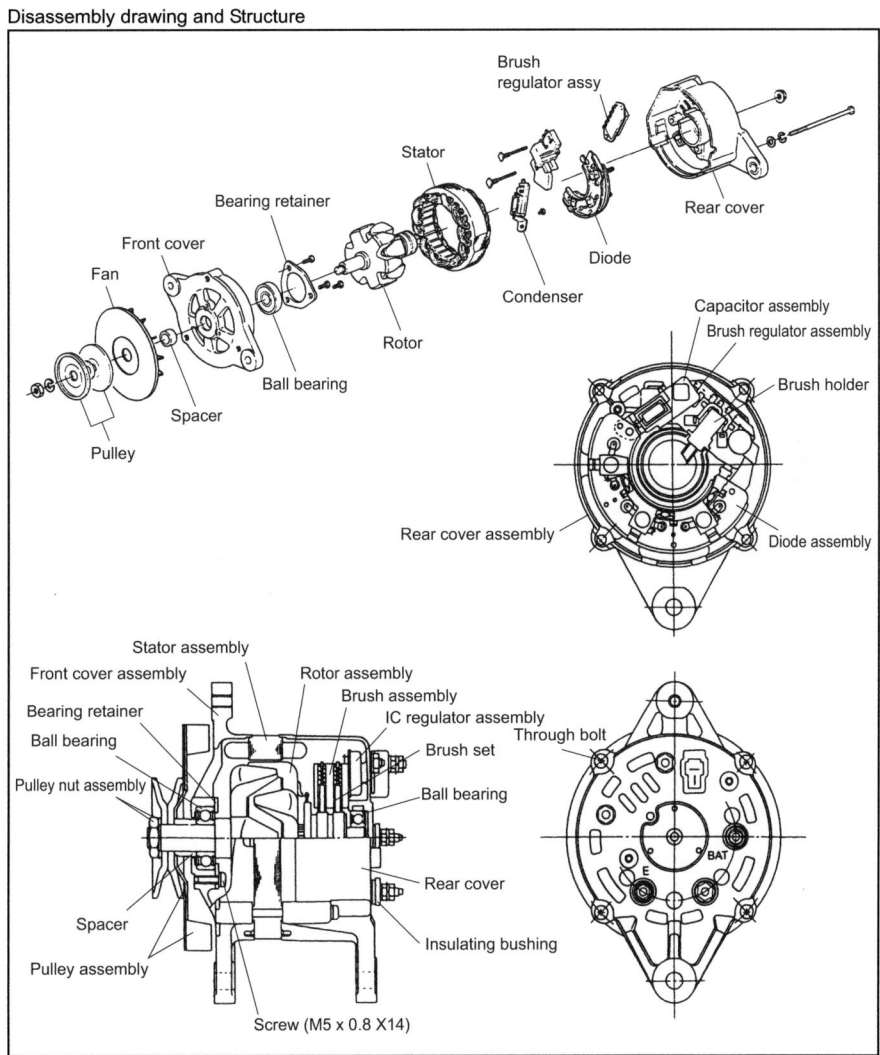

12.5.3 Wiring diagram

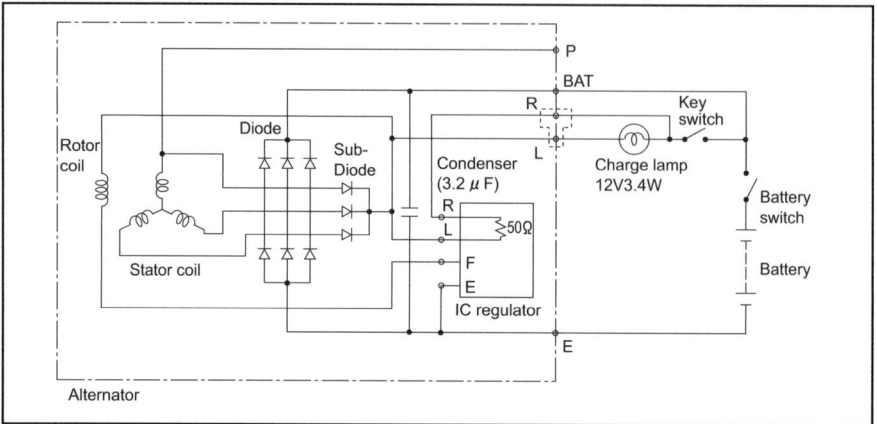

[NOTICE]
1) Don't do mis-connecting and short-circuit of each terminal.
2) Don't remove a battery terminal and a B terminal when rotating.
3) Shut out a battery switch during the alternator stop.

12.5.4 Standard output characteristics

The standard output characteristics of this alternator are shown as the below figure.

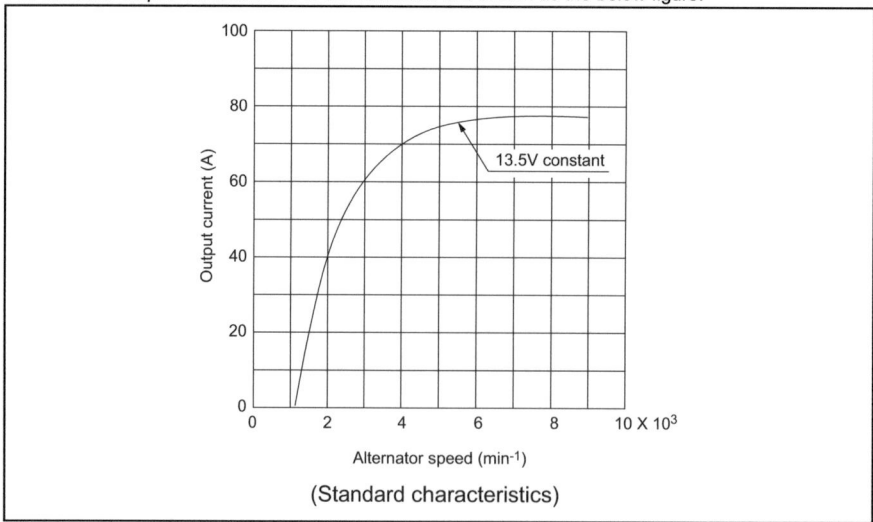

(Standard characteristics)

12.6 Instrument panel

The new type instrument panels are applied for 3YM30/3YM20/2YM15 series engines. The features are compactness, waterproof and independence from pulse by ring gear teeth number.
The engine speed, indicated with the instrument panel is activated by the pulse from flywheel ring gear.
The engine speed with new panel is activated by alternator B terminal pulse

12.6.1 B-type instrument panel (Optional)

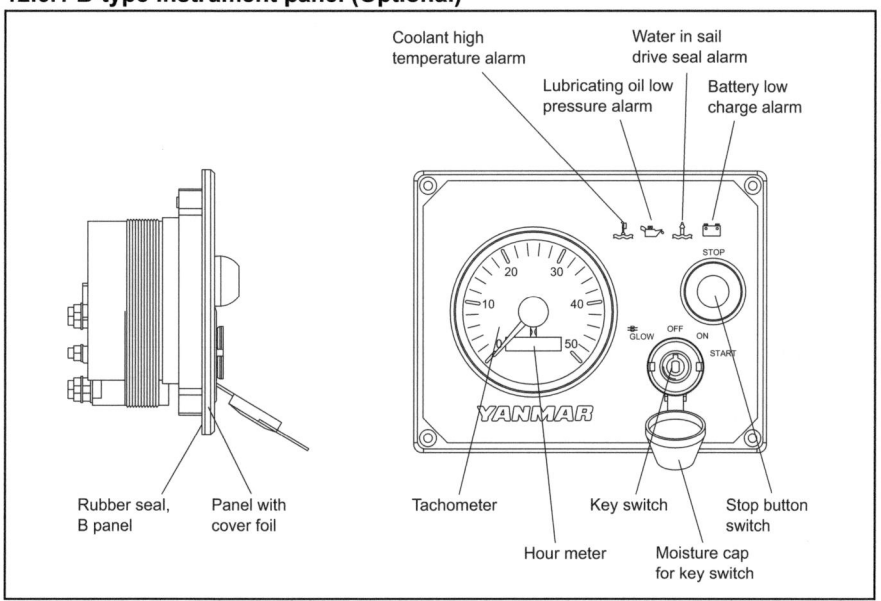

12.7 Warning devices

12.7.1 Oil pressure alarm

If the engine oil pressure is below 0.01-0.03 MPa (0.1-0.3 kgf/cm^2, 1.42-4.26 lb/in.^{2}), with the main switch in the ON position, the contacts of the oil pressure are closed by a spring and the lamp is illuminated through the lamp → oil pressure switch → ground circuit system. If the oil pressure is normal, the contacts of the switch are opened by the lubricating oil pressure and the lamp remains off.

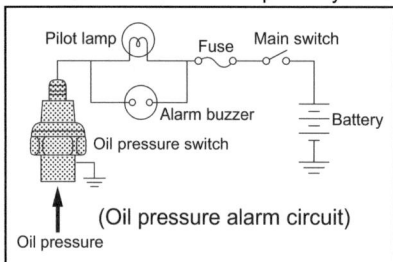

(Oil pressure alarm circuit)

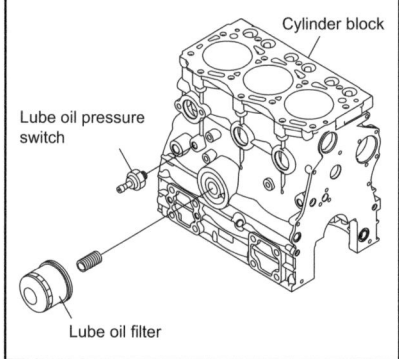

Rated voltage	12V
Operating pressure	0.04-0.06 MPa (0.4-0.6 kgf/cm^2)
Lamp capacity	5 W

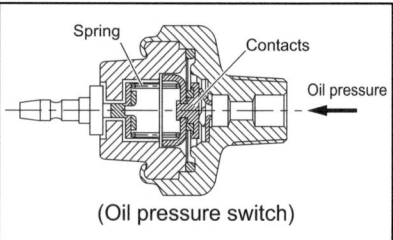

(Oil pressure switch)

12. Electrical System

Inspection

Problem	Inspection Item	Inspection method	Corrective action
Lamp not illuminated when main switch set to ON.	1. Oil pressure lamp blown out.	(1) Visual inspection. (2) Lamp not illuminated even when main switch set to ON position and terminals of oil pressure switch grounded.	Replace lamp.
	2. Operation of oil pressure switch.	Lamp illuminated when checked as described in (2) above.	Replace oil pressure switch.
Lamp not extinguished while engine running.	1. Oil level low.	Stop engine and check oil level with dipstick.	Add oil.
	2. Oil pressure low.	Measure oil pressure.	Repair bearing wear and adjust regulator valve.
	3. Oil pressure faulty.	Switch faulty if abnormal at (1) and (2) above.	Replace oil pressure switch.
	4. Wiring between lamp and oil pressure switch faulty.	Cut the wiring between the lamp and switch and wire with separate wire.	Repair wiring harness.

12.7.2 Cooling water temperature alarm

A water temperature lamp and a water temperature switch (thermo switch), backed up by an alarm in the instrument panel, are used to monitor the temperature of the engine cooling water. A high thermal expansion material is set on the end of the water temperature unit. When the cooling water temperature reaches a specified high temperature, the contacts are closed, and an alarm lamp and buzzer are activated at the instrument panel.

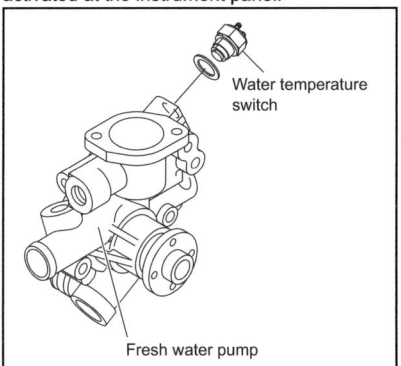

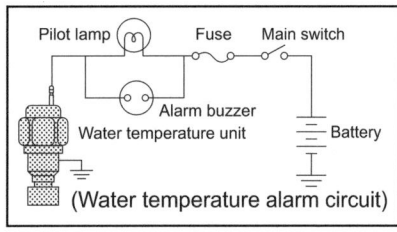

(Water temperature alarm circuit)

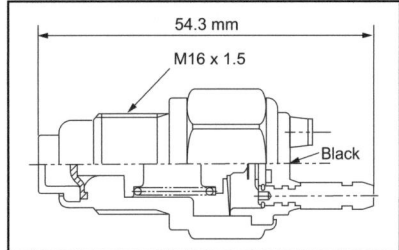

Operating temperature	ON	93-97 deg. C
Electric capacity		DC 12V, 1A
Response time		within 60 sec.
I.D. color		Black
Tightening torque		23.54-31.38 N·m (2.40-3.20 kgf-m)

12.8 Glow plug

A glow plug is available for warming intake air when starting in cold areas in winter. The glow plug is mounted to the cylinder head. The device is operated by the glow switch on the instrument panel.

Rated current	8-10A
Rated voltage	DC11V

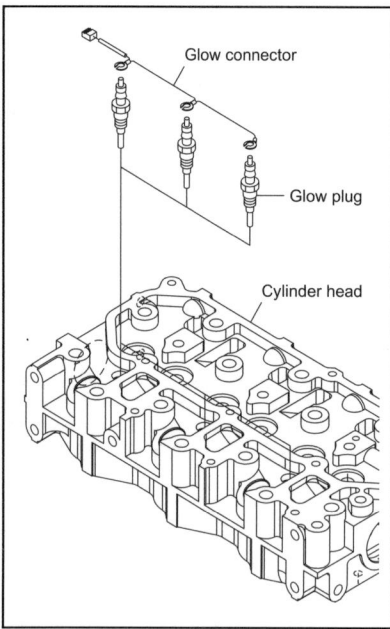
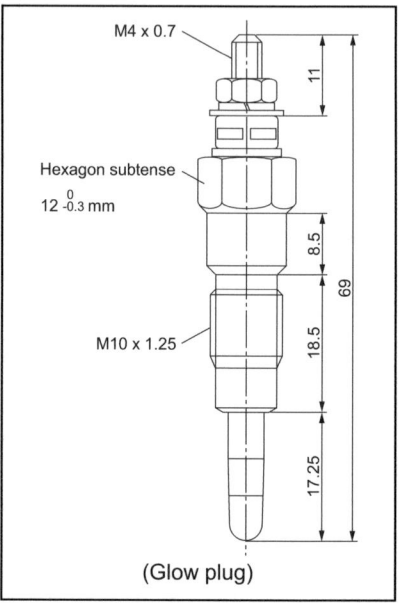

(Glow plug)

12.9 Electric engine stopping device

The electric engine stop device is fitted to the governor. The device is operated by the stop switch on the instrument panel.
The emergency stop button is integrated with the solenoid. When pushing the stop button, the engine will shut down.

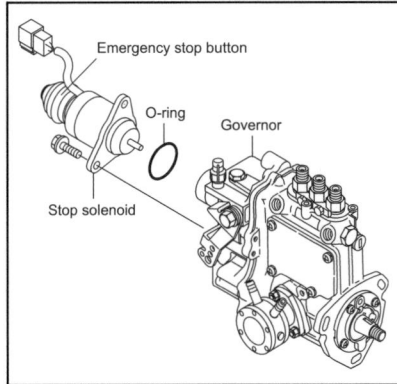

13. Service standards

13.1 Engine tuning

No.	Inspection item			Standard	Limit	Reference page
1	Intake/exhaust valve clearance mm			0.15-0.25	-	2.2.2(5)
2	V-belt tension at 98N (10kgf) mm	Between alternator and F.W. pump	Used part	8-10	-	2.2.2(4)
			New part	6-8	-	
3	Fuel injection pressure MPa (kgf/cm^2)			12.3-13.28 (125-135)	-	2.2.7(2)
4	Compression pressure (at 250 min^{-1}) MPa (kgf/cm^2)		3YM30	3.43±0.1 (35±1)	2.75±0.1 (28±1)	3.4
			3YM20 2YM15	3.23±0.1 (33±1)	2.55±0.1 (26±1)	
5	Cooling water Capacity Liter (quart)	Engine	3YM30	4.9 (5.2)	-	2.2.1(5)
			3YM20	4.1 (4.3)	-	
			2YM15	3.0 (3.2)	-	
		Coolant recovery tank		0.8 (0.8)	-	
6	Lube oil capacity (full)	3YM30 Liter (quart) (at rake angle 8 degree)		2.8 $^0/_{-0.2}$ (3.0)	-	2.2.2.(2) 2.2.2.(3)
		3YM20 Liter (quart) (at rake angle 8 degree)		2.7 $^0/_{-0.2}$ (2.9)	-	
		2YM15 Liter (quart) (at rake angle 8 degree)		2.0 $^0/_{-0.2}$ (2.1)	-	
		Marine gear KM2P-1 Liter (pint)		0.30 (0.64)	-	
		3YM30C Liter (quart) (at rake angle 0 degree)		2.5 $^0/_{-0.2}$ (2.6)	-	
		3YM20C Liter (quart) (at rake angle 0 degree)		2.4 $^0/_{-0.2}$ (2.5)	-	
		2YM15C Liter (quart) (at rake angle 0 degree)		1.8 $^0/_{-0.2}$ (1.9)	-	
		Sail drive SD20 Liter (pint)		2.2 (4.7)	-	
7	Lubricating oil pressure MPa (kgf/cm^2)	at rated speed		0.29-0.44 (3.0-4.5)	-	-
		at low idle speed		0.06(0.6) or above	-	8.2.5
8	Oil pressure switch operating pressure MPa (kgf/cm^2)			0.05±0.01 (0.5±0.1)	-	12.7.1
9	Thermostat	valve opening temperature deg. C		69.5-72.5	-	2.5
		Full opening lift (mm) (temperature)		8 or above (85 deg. C)	-	
10	Thermo switch actuating temperature (deg. C)	ON		93-97	-	2.4.2 12.7.2

13.2 Engine body

13.2.1 Cylinder head

(1) Cylinder head

Inspection item		Standard	Limit	Reference page
Combustion surface distortion mm		0.05 or less	0.15	5.2.1(1)
Valve sink mm	Intake	0.4-0.6	0.8	5.2.3(3)
	Exhaust			
Valve seat	Seat angle deg.	Intake	120	-	5.2.1(3)
		Exhaust	90	-	

(2) Intake/exhaust valve and guide

mm

Inspection item		Standard	Limit	Reference page
Intake	Guide inside diameter	6.000-6.012	6.08	5.2.3
	Valve stem outside diameter	5.960-5.975	5.90	
	Clearance	0.025-0.052	0.16	
Exhaust	Guide inside diameter	6.000-6.012	6.08	
	Valve stem outside diameter	5.945-5.960	5.90	
	Clearance	0.040-0.067	0.17	
Valve guide projection from cylinder head		9.8-10.0	-	5.2.3(4)
Valve guide driving-in method		Cold-fitted	-	

(3) Valve spring

mm

Inspection item	Standard	Limit	Reference page
Free length	37.8	36.3	5.2.4(1)
Inclination	-	1.3	

(4) Rocker arm and shaft

mm

Inspection item	Standard	Limit	Reference page
Arm shaft hole diameter	12.000-12.020	12.07	5.2.7(1)
Shaft outside diameter	11.966-11.984	11.94	
Clearance	0.016-0.054	0.13	

(5) Tappet and push rod

mm

Inspection item	Standard	Limit	Reference page
Tappet outside diameter	20.927-20.960	20.907	5.6.2(2)
Tappet guide hole inside diameter (cylinder block)	21.000-21.021	21.041	
Tappet oil clearance	0.040-0.094	0.134	
Push rod bend	Less than 0.03	0.03	5.6.2(3)

13.2.2 Camshaft and gear train

(1) Camshaft

mm

Inspection item			Standard	Limit	Reference page
Side gap			0.05-0.15	0.25	5.6.1(1)
Bending (1/2 the dial gage reading)			0.02 or less	0.05	5.6.1(4)
Cam height	3YM30		34.135-34.265	33.89	5.6.1(2)
	3YM20/2YM15		34.535-34.665	34.29	
Shaft outside diameter / Metal inside diameter					
Gear side	Bushing inside diameter		40.000-40.075	40.150	5.6.1(3)
	Camshaft outside diameter		39.940-39.960	39.905	
	Clearance		0.040-0.135	0.245	
Intermediate	Bushing inside diameter		40.000-40.025	40.100	
	Camshaft outside diameter		39.910-39.935	39.875	
	Clearance		0.065-0.115	0.225	
Flywheel side	Bushing inside diameter		40.000-40.025	40.100	
	Camshaft outside diameter		39.940-39.960	39.905	
	Clearance		0.04-0.085	0.195	

(2) Idle gear shaft and bushing

mm

Inspection item	Standard	Limit	Reference page
Shaft outside diameter	36.950-36.975	36.900	5.7.1(3)
Bushing inside diameter	37.000-37.025	37.075	
Clearance	0.025-0.075	0.175	

(3) Backlash of each gear

mm

Inspection item	Standard	Limit	Reference page
Crank gear, cam gear, idle gear and fuel injection pump gear	0.06-0.12	0.14	5.7.1(1)

13. Service standards

13.2.3 Cylinder block

(1) Cylinder block

mm

Inspection item		Standard	Limit	Reference page
Cylinder inside diameter	3YM30	76.000-76.030	76.200	5.1.5
	3YM20/2YM15	70.000-70.030	70.200	
Cylinder bore	Roundness	0.01 or less	0.03	
	Inclination			

(2) Crankshaft

mm

Inspection item		Standard	Limit	Reference page
Bending (1/2 the dial gauge reading)		-	0.01	5.5.1(2)
Crank pin 3YM30	Pin outside diameter	41.952-41.962	41.902	5.4.2(1) 5.5.1(3)
	Metal inside diameter	41.982-42.010	-	
	Metal thickness	1.503-1.509	-	
	Clearance	0.020-0.058	0.120	
Crank pin 3YM20/2YM15	Pin outside diameter	37.952-37.962	37.902	
	Metal inside diameter	37.982-38.010	-	
	Metal thickness	1.503-1.509	-	
	Clearance	0.020-0.058	0.120	
Crank journal (Selective pairing) All models	Journal outside diameter	46.952-46.962	46.902	
	Metal inside diameter	46.982-47.002	-	
	Metal thickness	2.009-2.014	-	
	Clearance	0.020-0.050	0.120	

(3) Thrust bearing

mm

Inspection item	Standard	Limit	Reference page
Crankshaft side gap	0.111-0.250	0.30	5.5.1(4)

13. Service standards

(4) Piston and ring

1) Piston

mm

Inspection item		Standard	Limit	Reference page
Piston outside diameter (Measure in the direction vertical to the piston pin.)	3YM30	75.965-75.975	75.920	5.3.1(1)
	3YM20/2YM15	69.970-69.980	69.925	
Piston diameter measure position (Upward from the bottom end of the piston)		22	-	
Clearance between piston and cylinder	3YM30	0.035-0.055	-	
	3YM20/2YM15	0.030-0.050	-	
Piston pin hole inside diameter		22.000-22.009	22.039	5.3.2
Piston pin outside diameter		21.995-22.000	21.965	
Clearance		0.000-0.014	0.074	

2) Piston ring

3YM30

mm

Inspection item		Standard	Limit	Reference page
Top ring	Groove width	1.550-1.570	-	5.3.3(1)
	Ring width	1.470-1.490	1.450	
	Side clearance	0.060-0.100	-	
	Ring gap	0.15-0.30	0.390	
Second ring	Groove width	1.580-1.595	1.695	
	Ring width	1.430-1.450	1.410	
	Side clearance	0.013-0.165	0.285	
	Ring gap	0.18-0.33	0.420	
Oil ring	Groove width	3.010-3.030	3.130	
	Ring width	2.970-2.990	2.950	
	Side clearance	0.020-0.060	0.180	
	Ring gap	0.20-0.45	0.540	

13. Service standards

3YM20/2YM15   mm

Inspection item		Standard	Limit	Reference page
Top ring	Groove width	1.550-1.570	-	5.3.3(1)
	Ring width	1.470-1.490	1.450	
	Side clearance	0.060-0.100	-	
	Ring gap	0.15-0.30	0.390	
Second ring	Groove width	1.540-1.560	1.660	
	Ring width	1.470-1.490	1.450	
	Side clearance	0.050-0.090	0.210	
	Ring gap	0.18-0.33	0.420	
Oil ring	Groove width	3.010-3.030	3.130	
	Ring width	2.970-3.010	2.950	
	Side clearance	0.020-0.060	0.180	
	Ring gap	0.15-0.35	0.44	

(5) Connecting rod
1) Rod big end

mm

Inspection item	Standard	Limit	Reference page
Side clearance	0.20-0.40	0.55	5.4.1(2)

2) Rod small end

mm

Item	Standard	Limit	Reference page
Bushing inside diameter	22.025-22.038	22.068	5.4.3(1)
Pin outside diameter	21.991-22.000	21.963	
Clearance	0.025-0.047	0.105	

13.3 Lubricating oil system (Trochoid pump)

(1) Outside clearance of outer rotor

mm

Standard	Limit	Reference page
0.12-0.21	0.30	8.2.4(1)

(2) Tip clearance between outer rotor and inner rotor

mm

Standard	Limit	Reference page
-	0.16	8.2.4(1)

(3) Side clearance of outer rotor

mm

Standard	Limit	Reference page
0.02-0.07	0.12	8.2.4(2)

(4) Outside diameter clearance of inner rotor centering location part

mm

Inspection item	Standard	Limit	Reference page
Gear case cover I.D.	46.13-46.18	-	8.2.4(3)
Inner rotor O.D.	45.98-46.00	-	
Rotor clearance	0.13-0.20	0.25	

14. Tightening torque for bolts and nuts

14.1 Main bolt and nut

No	Name	Thread diameter x pitch	Lube oil application (thread portion and seat surface)	Torque N·m(kgf·m)
1	Head bolt	M9 x 1.25	Coat with lube oil	53.9-57.9 (5.5-5.9)
2	Rod bolt	M7 x 1.0	Coat with lube oil	22.6-27.5 (2.3-2.8)
3	Flywheel retainer bolt	M10 x 1.25	Coat with lube oil	80.4-86.4 (8.2-8.8)
4	Metal cap retainer bolt	M10 x 1.25	Coat with lube oil	75.5-81.5 (7.7-8.3)
5	Crankshaft pulley bolt (FC250 pulley)	M10 x 1.25	Coat with lube oil	83.3-93.3 (8.5-9.5)
6	Fuel pump gear nut	M12 x 1.75	Coat with lube oil	58.8-68.8 (6.0-7.0)
7	Nozzle fastening nut	M20 x 1.5	No lube oil	49-53 (5.0-5.4)
8	Fuel injection pipe joint nut	M12 x 1.25	No lube oil	29.4-34.4 (3.0-3.5)
9	Glow plug	M10 x 1.25	No lube oil	14.7-19.6(1.5-2.0)
10	Governor weight support fastening nut	M12 x 1.25	Coat with lube oil	68.7-73.7(7.0-7.5)

14.2 Standard bolts and nuts (without lube oil)

N·m (kgf·m)

Name	Screw dia. x pitch	Tightening torque	Remarks
Hexagon bolt (7T) and nut	M6 x 1	9.8-11.8(1.0-1.2)	Use 80% of the value at left when the tightening part is aluminum.
	M8 x 1.25	22.5-28.5(2.3-2.9)	
	M10 x 1.5	44-54(4.5-5.5)	
	M12 x 1.75	78.2-98.2(8.0-10.0)	Use 60% of the value at left for 4T bolts and lock nuts.
PT plug	1/8	9.8 (1.0)	-
	1/4	19.6 (2.0)	
	3/8	29.4 (3.0)	
	1/2	58.8 (6.0)	
Pipe joint bolt	M8	12.7-16.7(1.3-1.7)	-
	M10	19.5-25.5 (2.0-2.6)	
	M12	24.4-34.4 (2.5-3.5)	
	M14	39.1-49.1 (4.0-5.0)	
	M16	48.9-58.9 (5.0-6.0)	